Scotland's leading educational publishers

Practice Papers for SQA Exams

Intermediate 2

Biology

ISBN 978-1-84372-776-7

Published by
Leckie & Leckie Ltd, 4 Queen Street, Edinburgh, EH2 1JE
Tel: 0131 220 6831 Fax: 0131 225 9987
enquiries@leckieandleckie.co.uk   www.leckieandleckie.co.uk

A CIP Catalogue record for this book is available from the British Library.

Leckie & Leckie Ltd is a division of Huveaux plc.

Questions and answers in this book do not emanate from SQA. All of our entirely new and original Practice Papers have been written by experienced authors working directly for the publisher.

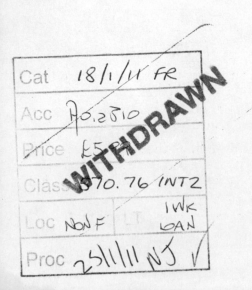

Introduction

Layout of the Book

This book contains practice exam papers, which mirror the actual SQA exam as much as possible. The layout, paper colour and question level are all similar to the actual exam that you will sit, so that you are familiar with what the exam paper will look like.

The answer section is at the back of the book. Correct answers are provided and, if appropriate, worked out solutions so that you can see how the right answers have been arrived at. The answers also include practical hints and tips on how to tackle certain types of questions, details of how marks are awarded and advice on just what the examiners will be looking for.

Revision advice is provided in this introductory section of the book, so please read on!

How To Use This Book

The Practice Papers can be used in two main ways:

1. You can complete an entire practice paper as preparation for the final exam. If you would like to use the book in this way, you should complete the practice paper under exam style conditions by setting yourself 2 hours for the paper and answering it as well as possible without using any references or notes. Alternatively, you can answer the practice paper questions as a revision exercise, using your notes to produce a model answer. Your teacher may mark these for you or you can mark your own answers using the answer section.
   A multiple choice grid is included that you may wish to photocopy and use as you would in the exam.

2. You can use the Topic Index at the front of this book to find all the questions within the book that deal with a specific unit. This allows you to focus specifically on areas that you particularly want to revise or, if you are part-way through your course, it lets you practise answering exam-style questions for individual units. Allow one minute per mark for any questions that you choose.

Revision Advice

Ideally, start many weeks before the exam dates but, it's never too late to start.

Work out a revision timetable for each week's work in advance – remember to cover all of your subjects and to leave time for homework and breaks. For example: early each evening do 30 minutes of normal homework, have a break then start your timetabled topics. If you have to miss one night, make sure these subjects get extra time the next week. Do not waste time redrafting your plan.

Day	6pm–6.45pm	7pm–8pm	8.15pm–9pm	9.15pm–10pm
Monday	Homework	Homework	English revision	Chemistry Revision
Tuesday	Maths Revision	Biology revision	Homework	Free
Wednesday	Geography Revision	Modern Studies Revision	English Revision	French Revision
Thursday	Homework	Maths Revision	Chemistry Revision	Free
Friday	Geography Revision	French Revision	Free	Free
Saturday	Free	Free	Free	Free
Sunday	Biology Revision	Maths Revision	Modern Studies	Homework

Make sure that you have at least one evening free a week to relax, socialise and re-charge your batteries. It also gives your brain a chance to process the information that you have been feeding it all week.

Arrange your study time into one hour or 30 minute sessions, with a break between sessions e.g. 6pm–7pm, 7.15pm–7.45pm, 8pm–9pm. Try to start studying as early as possible in the evening when your brain is still alert and be aware that the longer you put off starting, the harder it will be to start!

Study a different subject in each session, except for the day before an exam.

Do something different during your breaks between study sessions – take a walk, have a cup of tea, or listen to some music. Don't let your 15 minutes expand into 20 or 25 minutes though!

Have your class notes and any textbooks available for your revision to hand as well as plenty of blank paper, a pen, etc. You may like to make key term sheets like the biology example below:

Key term	Note
renal artery	blood vessel that carries blood into kidney
renal vein	blood vessel that carries blood out of kidney
ADH	hormone made by pituitary, increases permeability of tubule and increases water reabsorption

Finally ignore all or some of the revision advice in this section if you are happy with your present way of studying. Everyone revises differently, so find a way that works for you!

Arrangements document.

You can obtain your own copy of the SQA Arrangements document which tells students, teachers and examiners what is to be examined. It states very clearly what you need to know. The information is arranged as three columns and the examination is based on the first two called CONTENT and NOTES.

You can obtain your own copy from the SQA website in the following way:

- Go to www.sqa.org.uk
- Click on pupil
- Click on Biology
- Click on Intermediate 2
- Click on Arrangements Documents
- Print out pages 6 to 28. Don't print out the whole publication as it is 64 pages long!

Transfer Your Knowledge

As well as using your class notes and textbooks to revise, these practice papers will also be a useful revision tool as they will help you to get used to answering exam style questions. You may find as you work through the questions that they refer to an investigation or an example that you haven't come across before. Don't worry! You should be able to transfer your problem solving skills or knowledge of a topic to a new example. Some questions are deliberately written to be unfamiliar to everyone sitting the exam. Read all the information provided e.g. describing how an experiment was carried out before you attempt to answer the questions.

Command Words

In the practice papers and in the exam itself, a number of command words will be used in the questions. These command words are used to show you how you should answer a question – some words indicate that you should write more than others. If you familiarise yourself with these command words, it will help you to structure your answers more effectively.

Command Words	Meaning/Explanation
Name, state, identify, what is, which term	Short answers are best to answer these knowledge and understanding questions – as a general rule you will get one mark for each point you give
Suggest	Give more than a list – perhaps a proposal or an idea from your biology knowledge
Describe, outline	Give details of what happens in processes or give functions using named examples
Explain	Give reasons **why** a conclusion has been reached or **why** an action has been taken
Compare	Give the key features of 2 different processes or ideas and discuss their similarities and/or their differences.
Define	Give the meaning of the term

Key Problem Solving Questions.

Problem Solving Skill	Explanation
Calculate	Use figures given or extracted from sources to work out averages, percentages or ratios
Predict	Look at the changes, pattern or trend of results to identify an earlier or later result
Draw a Conclusion	Identify the pattern or trend of results related to the aim of the investigation.
Valid results	Results obtained from an investigation in which all variables are kept constant except one. Remember validity = variables.
Reliable results	Results obtained when an investigation has been repeated and results compared. Remember reliability = repeats.

In the Exam

Watch your time and pace yourself carefully. Work out roughly how much time you can spend on each section and try to keep to this. If you are stuck, move on.

Be clear before the exam what the instructions are likely to be e.g. using the separate grid to answer section A questions and choosing two extended questions to do in section C. The practice papers will help you to become familiar with the exam's instructions.

You can tackle the questions in any order so, as you start section B, look through and find a question that you will do well to answer first.

Read the question thoroughly before you begin to answer it – make sure you know exactly what the question is asking you to do e.g. if you are told to draw a line graph, do not draw a bar graph.

In section C, plan each answer by jotting down keywords, a mindmap or reminders of the important things to include. Cross them off as you deal with them and check them before you move on to the next question to make sure that you haven't forgotten anything.

Don't repeat yourself as you will not get any more marks for saying the same thing twice. This also applies to annotated diagrams which will not get you any extra marks if the information is repeated in the written part of your answer. If you have to give ways of improving validity of results, do not restate the validity points given in the experiment description, think out others that should apply.

Give proper explanations. A common error is to give descriptions rather than explanations. If you are asked to explain something, you should be giving reasons. Check your answer to an 'explain' question and make sure that you have used plenty of linking words and phrases such as 'because', 'this means that', 'therefore', 'so', 'so that', 'due to', 'since' and 'the reason is'.

Other common exam errors include

- drawing a diagram and not putting in the necessary labels
- using the word amount when you should say volume, concentration, mass or number
- using the terms like or prefer when referring to behavioural responses
- forgetting graph labels or units

Use the resources provided. Some questions will ask you to 'describe and explain' and provide a graph or table of data for you to work from. Make sure that you take any relevant data from these resources.

Good luck!

Topic Index

Topic		Exam 1	Exam 2	Exam 3
Unit 1 Living Cells				
Cells	A	1, 2	1	1, 2
	B		1a	5
	C			
Uses of cells	A	3, 4	2	3
	B		1b(i)	5
	C	2B		
Osmosis	A		4	4, 5, 6
	B	1		
	C			
Diffusion	A	5	3	
	B			
	C			
Enzymes	A	6, 7, 8	5, 6, 7	
	B	2	2	6
	C			
Respiration	A		8	
	B	3	3, 1(b)(ii)	7
	C			
Photosynthesis	A		9	7, 8
	B	4	4	8
	C	2A		
Unit 2 Environmental Biology and Genetics				
Ecosystems	A	16	17	9–13, 14, 15
	B	5, 8		2B
	C			
Behaviour	A		5 6(a)/(b)	14
	B			
	C			
Fertilisation	A	11, 15		
	B			1
	C		1A	
Chromosomes and DNA	A	13	15	
	B			2
	C			
Gamete Formation – Meiosis	A	14, 17		
	B			1
	C		1A	
Genes and Alleles	A	6(a)	10	
	B			
	C			

(continued)

Topic (Continued)		Exam 1	Exam 2	Exam 3
Monohybrid Crosses	A		11, 12, 13, 14	
	B	6(b)		3
	C			
Inheritance	A		16	
	B			3
	C			
Natural selection	A	10		
	B			4
	C		1B	
Selective Breeding	A	9		
	B		8	
	C			
Genetic Engineering	A	12		
	B		7	
	C			2A
Unit 3 Animal Physiology				
Food	A			24, 25
	B		11	9, 12
	C			
Digestion	A	19, 20	22	
	B		9	9
	C	1B		
Kidneys	A	21	19, 20	
	B	11		
	C			1B
Osmoregulation	A		18, 21	
	B	11		
	C	1A		
Circulatory system	A	22	23, 24, 25	17–20
	B	12		10, 11
	C		2B	
Lungs	A			
	B	9	12	11
	C			1A
Nervous system	A	24, 25		21, 22, 23
	B		10	
	C			
Temperatur regulation	A			
	B	10		
	C		2A	

Practice Exam 1

Biology   Intermediate 2

Practice Papers
For SQA Exams

Exam 1
Intermediate 2

Fill in these boxes:

Name of centre

Town

Forename(s)

Surname

You have 2 hours to complete this paper.

Instructions

Section A (25 marks)

- Complete the grid provided.

- Full instructions on next page.

Section B and C (75 marks)

- Attempt all questions and try to leave no spaces.

- Section C – two questions. Choose one option from each.

- Answer questions in ink and in any order.

- Complete graphs in pencil.

Scotland's leading educational publishers

Indicate your choice of answer with a single
mark in pencil as in the following example. ⟶

	A	B	C	D
	☐	■	☐	☐

	A	B	C	D			A	B	C	D
1	☐	☐	☐	☐		14	☐	☐	☐	☐
2	☐	☐	☐	☐		15	☐	☐	☐	☐
3	☐	☐	☐	☐		16	☐	☐	☐	☐
4	☐	☐	☐	☐		17	☐	☐	☐	☐
5	☐	☐	☐	☐		18	☐	☐	☐	☐
6	☐	☐	☐	☐		19	☐	☐	☐	☐
7	☐	☐	☐	☐		20	☐	☐	☐	☐
8	☐	☐	☐	☐		21	☐	☐	☐	☐
9	☐	☐	☐	☐		22	☐	☐	☐	☐
10	☐	☐	☐	☐		23	☐	☐	☐	☐
11	☐	☐	☐	☐		24	☐	☐	☐	☐
12	☐	☐	☐	☐		25	☐	☐	☐	☐
13	☐	☐	☐	☐						

Multiple Choice Instructions

- Check the separate grid provided and complete the details requested.

- Check that **your name** is on the answer grid.

- Read each question carefully and decide on your answer.

- Tick the correct answer on your question paper.

- Using **pencil**, transfer your answer to the separate grid.

- Shade in the box under your chosen answer.

To change an answer, erase your answer fully and shade in the new box.

Check your answers before you leave the exam room.

Leave no spaces – *all* questions should have an answer.

Make sure you only give one answer to each question.

SECTION A

Try to answer all of the questions in Section A.

Use the grid provided.

Remember to complete the grid in pencil.

1. Animal cells contain a

   A   nucleus and chloroplasts
   B   cell membrane and cell wall
   C   vacuole and chloroplasts
   D   nucleus and cytoplasm.

2. The function of a cell wall is to

   A   control entry and exit of materials
   B   support and give shape
   C   prevent plasmolysis
   D   control diffusion.

3. Which line in the table below is correct for antibiotics?

	Produced by	Effect on
A	yeast	fungi
B	lymphocytes	yeast
C	fungi	bacteria
D	microbial cells	fungi

4. Which line in the table below shows correctly the formation of biogas?

	Produced by	Form of Respiration	Biogas produced
A	bacteria	aerobic	gasohol
B	yeast	aerobic	methane
C	bacteria	anaerobic	methane
D	yeast	anaerobic	gasohol

5. Which line in the table below is correct for diffusion?

	Substance	Enters or Leaves Cell	Function
A	protein	leaves	cell product
B	oxygen	enters	waste product
C	carbon dioxide	leaves	photosynthesis
D	glucose	enters	respiration

6. Which of the following graphs, A, B, C or D represents the activity of amylase?

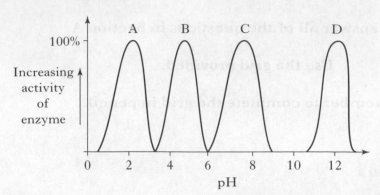

7. The diagrams below show stages in an enzyme-catalysed reaction.

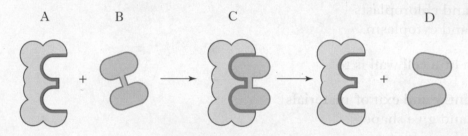

Which letter identifies the enzyme?

8. Which line in the table below correctly shows an enzyme, its substrate and its products?

	Enzyme	Substrate	Product
A	amylase	glucose	starch
B	catalase	hydrogen peroxide	carbon dioxide
C	phosphorylase	glucose-1-phosphate	cellulose
D	amylase	starch	maltose

Questions 9 and 10 refer to the answers below

    A   biodiversity
    B   natural selection
    C   selective breeding
    D   genetic engineering.

9. Which term is used to describe the process of choosing the best of each generation to produce the next generation?

10. Which term is used to describe the survival of the fittest in an environment?

11. In humans, which of the following gametes would result in a male offspring after fertilisation?

    A   an X egg and Y sperm.
    B   an Y egg and X sperm.
    C   an Y sperm and Y egg.
    D   an X sperm and an X egg.

12. The following stages occur during genetic engineering.

    1. plasmid and gene inserted into bacterial cell
    2. gene cut out of chromosome
    3. gene sealed into plasmid
    4. plasmid cut open.

    The correct sequence of these stages is

    A   1 2 4 3
    B   2 4 3 1
    C   1 2 3 4
    D   2 4 1 3.

13. Which of the following cells all contain two sets of chromosomes?

    A   sperm, eggs and zygotes
    B   macrophages, eggs and sperm
    C   lymphocytes, sperm and eggs
    D   zygotes, macrophages and lymphocytes

14. Each cell undergoing meiosis in the testes produces

    A   four different sperm
    B   two identical sperm
    C   four different zygotes
    D   two identical zygotes

15. Which of the following is a correct description of fertilization in the ovary of a plant?

    A   formation of ovules.
    B   two sets of genes are divided to form a zygote.
    C   the nuclei of gametes fuse to form a zygote.
    D   production of gametes

16. Desert plants do **not** save water by using

    A   succulent tissue
    B   spines instead of leaves
    C   waxy cuticle on leaves
    D   deep roots.

17. Which line below names a gamete and its site of production in a flowering plant?

    A   sperm produced in the testes.
    B   eggs produced in ovary.
    C   pollen produced in the testes.
    D   zygote produced in the ovary.

18. A chocolate bar weighs 20 g of which 6·5 g is fat.

What is the percentage fat in the chocolate bar?

A   3·1
B   13·0
C   30·7
D   32·5

19. The diagram shows a villus from the small intestine.

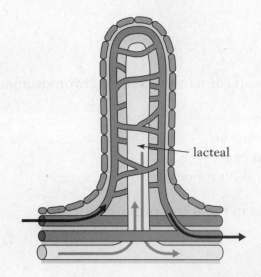

The lacteal absorbs

A   amino acids and glucose
B   amino acids and glycerol
C   fatty acids and glycerol
D   fatty acids and glucose.

20. Bile is stored in the

A   small intestine
B   gall bladder
C   large intestine
D   liver.

21. Human kidneys filter out 180 litres of liquid a day and 99% of the water is reabsorbed.

What volume of urine is passed out of the body?

A   1·8 litres
B   18 litres
C   162 litres
D   178·2 litres

22. Oxygen is transported by being combined with haemoglobin to form oxyhaemoglobin.

Which of the following statements is true?

A   At high oxygen levels, oxyhaemoglobin releases oxygen in the lungs.
B   At high oxygen levels, oxyhaemoglobin releases oxygen in the tissues.
C   At low oxygen levels, oxyhaemoglobin releases oxygen in the lungs.
D   At low oxygen levels, oxyhaemoglobin releases oxygen in the tissues.

23. Nationally, spending on anti-obesity drugs has increased over the last few years.

In 2000, annual spending was £12 600 and by 2008, the spending was £365 400.

By how many times has spending increased over this period?

A   8 times.
B   29 times.
C   44 100 times.
D   45 674 times.

Questions 24 and 25 refer to the diagram of the brain shown below.

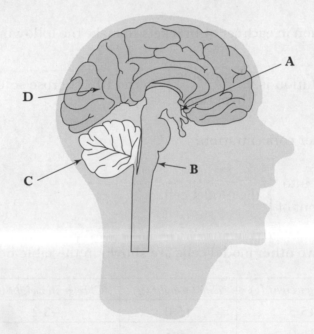

24. Which labelled structure is the cerebellum?

25. Which labelled structure detects changes in the water content of the blood?

SECTION B

Try to answer all of the questions in Section B.

1.   In an investigation into osmosis, a student set up the following apparatus.

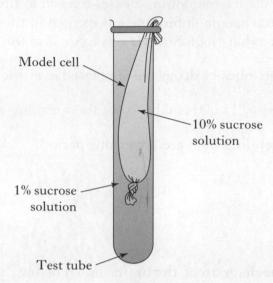

Model cell

10% sucrose
solution

1% sucrose
solution

Test tube

(a) <u>Underline</u> **one** option in each set of brackets to make the following sentences correct.

The 1% sucrose solution is $\left\{ \begin{array}{l} \text{hypotonic} \\ \text{hypertonic} \end{array} \right\}$ to the 10% sucrose solution which

has a $\left\{ \begin{array}{l} \text{higher} \\ \text{lower} \end{array} \right\}$ water concentration.

Water will move $\left\{ \begin{array}{l} \text{into} \\ \text{out of} \end{array} \right\}$ the model cell.   2

(b) The results from two other model cells are shown in the table below.

Model cell	Starting weight (g)	Final weight (g)	Change in weight (%)
1	15·5	15·0	−3·2
2	16·0	16·8	

(i) Complete the table to show the percentage change in weight for model cell 2.

Space for calculation

1

    (ii) Why is it necessary to calculate the **percentage change** in weight rather than the actual change in weight?

_____   1

    (iii) What name is given to a solution containing the same water concentration as a cell?

_____   1

2. In an investigation into the breakdown of hydrogen peroxide, a student set up three measuring cylinders as shown in the diagrams below.

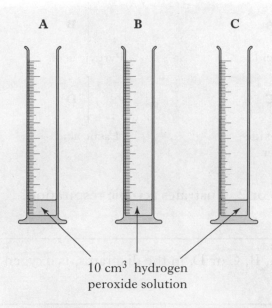

10 cm³ hydrogen
peroxide solution

The student added 2 cm³ of water to cylinder A, 2 cm³ of amylase solution to cylinder B and 2 cm³ catalase solution to cylinder C.

(*a*) (i) Predict what the student would observe in measuring cylinder B. Explain your answer.

Prediction _____   1

Explanation _____   1

    (ii) What was the purpose of measuring cylinder A?

_____   1

    (iii) Name the gas produced in measuring cylinder C.

_____   1

(b) How could the student measure the rate of enzyme's reaction in this investigation?

_____

_____ 2

3. The diagrams below show stages in respiration

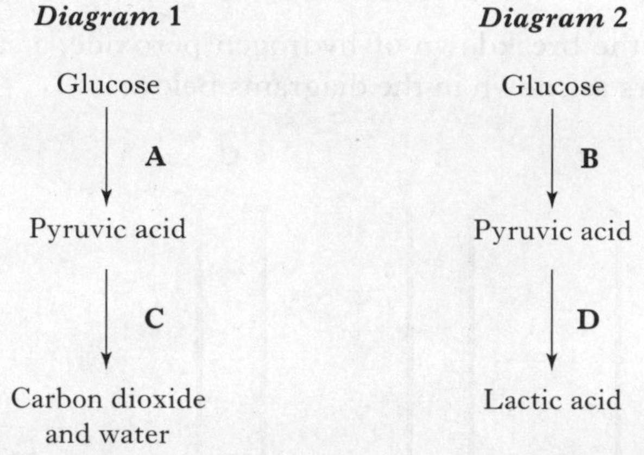

Diagram 1	*Diagram* 2
Glucose	Glucose
↓ A	↓ B
Pyruvic acid	Pyruvic acid
↓ C	↓ D
Carbon dioxide and water	Lactic acid

(a) (i) Which diagram, 1 or 2, illustrates aerobic respiration?

_____ 1

(ii) At which point, A, B, C or D, in the diagrams, is oxygen needed?

_____ 1

(b) Give **two** examples of the uses that the human body makes of the energy that is released from respiration.

Use 1 _____ 1

Use 2 _____ 1

4. The diagram below shows the first stage in photosynthesis.

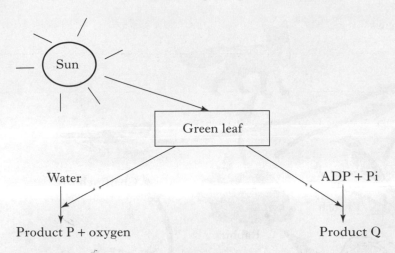

(a) Which substance in a green leaf traps the energy of sunlight?

_____ 1

(b) (i) Name the two products **P** and **Q**.

P _____ 1

Q _____ 1

(ii) Products P and Q enter the second stage of photosynthesis.
Name this second stage.

_____ 1

(iii) What happens to **product P** during this second stage of photosynthesis?

_____

_____ 1

5. The diagram below shows part of a woodland food web.

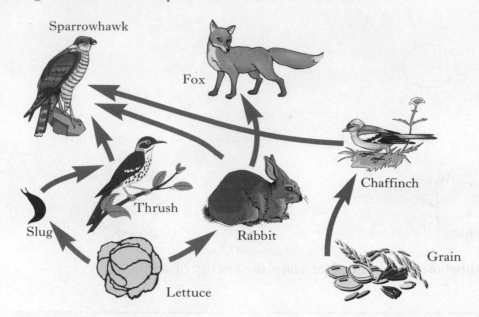

(*a*) Decide if each of the following statements about the food web is **True** or **False**, and tick (✓) the appropriate box.

If the statement is **False**, write the correct word in the **Correction** box to replace the word underlined in the statement.

Statement	True	False	Correction
The <u>fox</u> is a producer in this food web.			
The sparrowhawk is a <u>herbivore</u>.			
The <u>predator</u> of the thrush is the slug.			

3

(*b*) Predict the effect on the rabbit population if the number of sparrowhawks is reduced by poisoning. Explain your answer.

Prediction _____

Explanation _____

_____

1

(c) The following terms are used to describe ecosystems. Draw a line to connect each term to its correct definition.

Term   **Definition**

organisms plus environment

Community

place where an organism lives

Population

a group of different species

Ecosystem

a group of the same species

3

6. The inheritance of flower colour in snapdragons was investigated using a series of genetic crosses. In the first cross red flowered plants were crossed with white flowered plants. All the F$_1$ plants were pink as shown below.

(a) Using the symbols R(red) and W(white) for the alleles, complete the diagram to show the genotypes of the parents and their gametes.

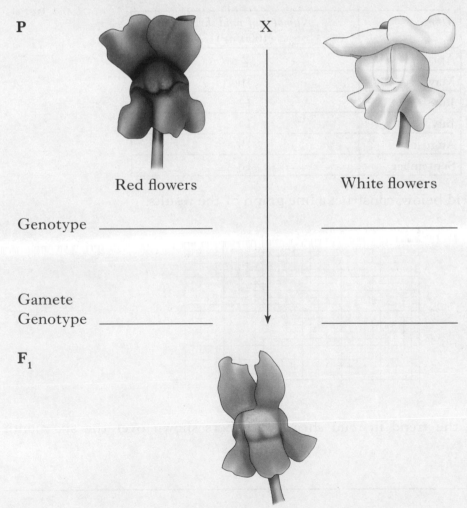

P   X

Red flowers   White flowers

Genotype _____   _____

Gamete
Genotype _____   _____

F$_1$

Pink flowers   2

(b) (i) An F_1 pink flowered plant is crossed with another F_1 pink flowered plant.

Complete the Punnett square below to show the expected results of this cross.

Gametes from pink parent 1

Gametes from pink parent 2

2

(ii) Give the expected phenotype ratio of the F_2 offspring from the cross between the two pink flowered plants.

_____ Red : _____ Pink : _____ White

1

7. The table below shows the number of mud shrimps found in an estuary each month over a six month period. These organisms are the prey of migratory birds which come to the estuary from August to April.

Month	Numbers of mud shrimp (1000/m²)
April	2
May	10
June	12
July	14
August	18
September	16

(a) On the grid below, construct a line graph of the results.

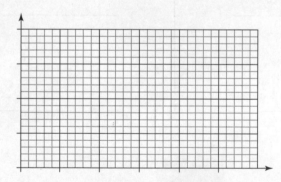

3

(b) Describe the trend in mud shrimp numbers shown over the six month period.

_____

_____

_____

2

(c) What is the average number of mud shrimps found per month?

Space for calculation

Average _____ 1000/m²   **1**

(d) (i) How many times greater are the numbers of mud shrimps in August compared to April?

Space for calculation

_____ times   **1**

(ii) From the information given, suggest a reason for this increase in numbers.

_____

_____   **1**

8. The graph below shows the changes that take place in a stretch of river polluted by sewage.

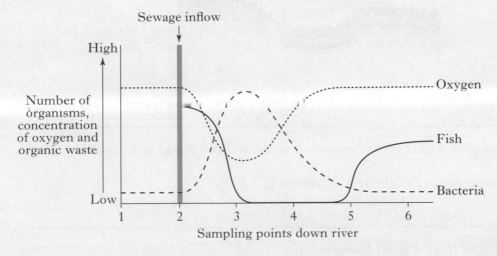

(a) Using information from the graph, state the immediate effect of adding sewage on the numbers of bacteria and oxygen concentration.

Bacteria _____

Oxygen Concentration _____   **1**

(b) Explain the effect of adding sewage on the number of fish found at point 4 on the graph.

_____

_____

_____ 2

(c) Pollution changes biodiversity in this river. Give the meaning of the term biodiversity.

_____ 1

(d) Give an example of a human activity, other than pollution, which might change biodiversity in this ecosystem.

_____ 1

9. (a) The diagram below represents an alveolus.

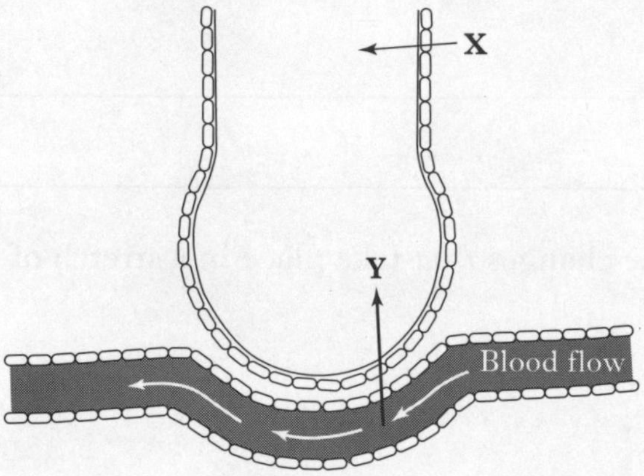

(i) Name structure X.

_____ 1

(ii) Gas exchange occurs between the air and the blood through the alveoli.

Name the gas indicated by arrow Y.

_____ 1

(iii) Alveoli have a good blood supply.

State two other features of alveoli that enable efficient gas exchange.

1. _____ 1

2. _____ 1

(b) The oxygen level in the blood circulating through two leg muscles was measured at rest and during exercise.

Leg muscle	Oxygen level of blood (%)	
	At rest	During exercise
P	70	63
Q	73·5	

(i) The oxygen level in muscle Q decreased during exercise.

The decrease was half as much as the decrease that occurred in muscle P.

Complete the table by calculating the oxygen level in muscle Q during exercise.

Space for calculation

1

(ii) Calculate the percentage decrease in oxygen level of the blood circulating through muscle P during exercise.

Space for calculation

_____ % reduction   1

(c) Blood flow to a muscle was measured at 800 cm³ per minute per kilogram of muscle tissue.

The mass of the muscle was 250g.

Calculate the volume of blood flowing through the muscle per minute.

Space for calculation

Volume _____ cm³.   1

10. The sentences below describe temperature regulation.

Underline **one** option in each set of brackets to make the sentences correct.

The temperature of the external environment is detected by $\begin{Bmatrix} \text{receptors} \\ \text{effectors} \end{Bmatrix}$ in the skin.

The part of the brain which regulates body temperature is the $\begin{Bmatrix} \text{cerebellum} \\ \text{hypothalamus} \end{Bmatrix}$.

An increase in the temperature of the surrounding environment causes arterioles in the skin to $\begin{Bmatrix} \text{dilate} \\ \text{constrict} \end{Bmatrix}$ and sweat production to $\begin{Bmatrix} \text{increase} \\ \text{decrease} \end{Bmatrix}$.

3

11. The diagram below represents methods of gain and loss of water from the human body.

Two examples are provided

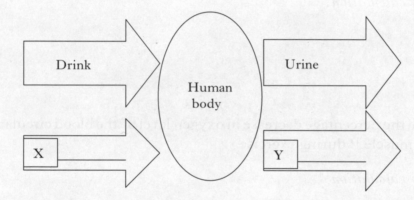

(a) Complete the arrows at X and Y to show one other gain and one other loss of water.

2

(b) One filtering unit from the kidneys is shown below.

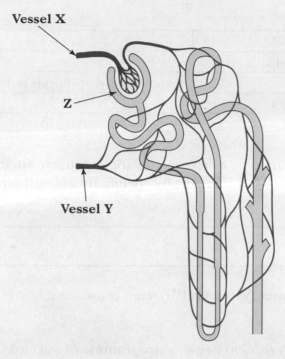

Vessel X

Z

Vessel Y

(i) Name structure Z.

_____ 1

(ii) Name the waste product present in vessel X that is not present in vessel Y.

_____ 1

(iii) A. Name the hormone that controls the volume of water reabsorbed.

_____ 1

B. From which part of the body is this hormone released?

_____ 1

12. (a) Complete the table below which describes some functions of parts of the blood.

Part of blood	Function
Red blood cells	
	Transports carbon dioxide and dissolved food
Macrophages	

2

(b) When the body is invaded by a disease-causing organism, specific antibodies are produced. Explain what is meant by specificity of antibodies.

_____

_____

1

(c) The Human Papilloma Virus (HPV) can cause cancer of the cervix in females.

In 2008 the UK Government began a programme of vaccinations that were offered to school age girls.

Figures from the programme of HPV vaccination carried out on girls in Scotland showed that by February 2009, 92% of fifth and sixth year girls had taken the first dose.

One school is identified as typical. It has 250 girls in this year grouping.

How many of them have been vaccinated against HPV?

Space for working

_____ girls   1

SECTION C

**Try to answer both questions in Section C,
but note that each question has a choice.**

Questions 1 and 2 should be attempted on seperate pages.

1. Answer **either** A **or** B.

    A. The tissues of a freshwater fish are hypertonic to its environment.

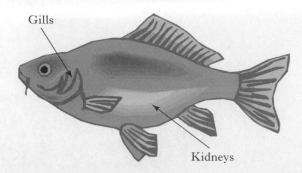

Gills

Kidneys

Describe how freshwater fish maintain their water balance. Your answer should mention water concentrations and the type of urine produced   **5**

    OR

    B. The diagram below shows a human stomach.

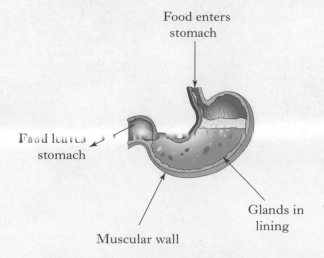

Food enters
stomach

Food leaves
stomach

Muscular wall

Glands in
lining

Describe the role of the stomach in digestion of food. Your answer should mention muscular movements and the types of cells in the stomach lining.   **5**

2. Answer **either** A **or** B.

   Labelled diagrams may be included where appropriate.

   A. Name the limiting factors in photosynthesis and explain how each of them can be overcome in a commercial greenhouse to increase profit.   5

   OR

   B. Single celled microorganisms are used in various industries. Name the microbial cells involved and how they are used in commercial production.   5

[End of Question Paper]

Practice Exam 2

Biology   Intermediate 2

Practice Papers   **Exam 2**
For SQA Exams   **Intermediate 2**

Fill in these boxes:

Name of centre

Town

Forename(s)

Surname

You have 2 hours to complete this paper.

Instructions

Section A (25 marks)

- Complete the grid provided.

- Full instructions on next page.

Section B and C (75 marks)

- Attempt all questions and try to leave no spaces.

- Section C – two questions. Choose one option from each.

- Answer questions in ink and in any order.

- Complete graphs in pencil.

Scotland's leading educational publishers

Indicate your choice of answer with a single mark in pencil as in the following example.

	A	B	C	D
→	☐	■	☐	☐

	A	B	C	D		A	B	C	D
1	☐	☐	☐	☐	14	☐	☐	☐	☐
2	☐	☐	☐	☐	15	☐	☐	☐	☐
3	☐	☐	☐	☐	16	☐	☐	☐	☐
4	☐	☐	☐	☐	17	☐	☐	☐	☐
5	☐	☐	☐	☐	18	☐	☐	☐	☐
6	☐	☐	☐	☐	19	☐	☐	☐	☐
7	☐	☐	☐	☐	20	☐	☐	☐	☐
8	☐	☐	☐	☐	21	☐	☐	☐	☐
9	☐	☐	☐	☐	22	☐	☐	☐	☐
10	☐	☐	☐	☐	23	☐	☐	☐	☐
11	☐	☐	☐	☐	24	☐	☐	☐	☐
12	☐	☐	☐	☐	25	☐	☐	☐	☐
13	☐	☐	☐	☐					

Multiple Choice Instructions

- Check the separate grid provided and complete the details requested.

- Check that **your name** is on the answer grid.

- Read each question carefully and decide on your answer.

- Tick the correct answer on your question paper.

- Using **pencil**, transfer your answer to the separate grid.

- Shade in the box under your chosen answer.

A B C D

To change an answer, erase your answer fully and shade in the new box.

Check your answers before you leave the exam room.

Leave no spaces – *all* questions should have an answer.

Make sure you only give one answer to each question.

SECTION A

Try to answer all of the questions in Section A.

Use the grid provided.

Remember to complete the grid in pencil.

1. Which line in the table below shows correctly cell structure and function?

	Cell structure	Function
A	nucleus	contains DNA
B	cytoplasm	controls cell activities
C	cell membrane	maintains cell shape
D	cell wall	controls entry and exit of materials

2. Which line in the table below shows correctly the commercial and industrial uses of cells?

	Cell type	Industry	Product
A	bacteria	baking	dough
B	yeast	brewing	antibiotics
C	bacteria	baking	carbon dioxide
D	yeast	brewing	ethanol

3. Which of the following is **not** an importance of diffusion to cells?

    A   uptake of water for photosynthesis
    B   removal of carbon dioxide produced during respiration
    C   removal of glucose produced during photosynthesis
    D   uptake of carbon dioxide for respiration

4. When an animal cell is placed into a hypertonic solution it will

    A   burst
    B   shrink
    C   become flaccid
    D   become turgid.

5. Which graph below shows correctly the effect of increasing temperature on enzyme action?

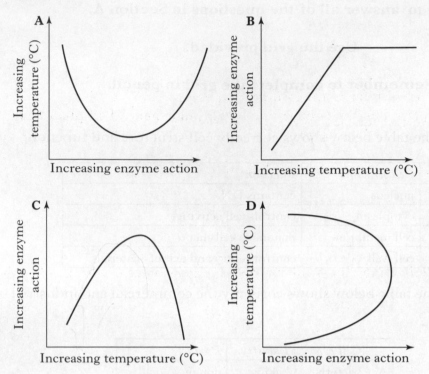

6. The enzyme amylase is specific to the substrate

    A   protein
    B   lipid
    C   starch
    D   maltose

7. The building up of large, complex molecules from smaller, simple molecules is called

    A   synthesis
    B   digestion
    C   degradation
    D   denaturation.

8. Respiration results in the

    A   overall uptake of energy
    B   formation of ADP and P_i
    C   release of heat energy
    D   conversion of pyruvic acid to glucose.

9. The benefit of supplementary lighting in a greenhouse is that it

    A   reduces the costs to the gardener
    B   allows earlier crops to be produced
    C   increases respiration in plants
    D   reduces the need for extra heating.

10. The term used to describe the appearance of an organism is

    A   genotype
    B   phenotype
    C   dominant
    D   recessive

    Questions 11 and 12 refer to the information below.
    In pea plants, plant height is determined by a single gene. Tall plant height (T)
    is dominant to dwarf plant height (t).

    A cross between two plants is shown below.

    P   TT   ×   tt

    F_1 Tt

11. Which of the following shows all the possible genotypes of the F_2 offspring if
    two F_1 plants are crossed?

    A   TT and Tt
    B   tt and Tt
    C   TT and tt
    D   TT, Tt and tt

12. What percentage of the F_2 generation would be expected to be dwarf plants?

    A   25%
    B   33%
    C   50%
    D   75%

13. When a tomato plant with a potato leaf was crossed with a tomato plant with a
    cut leaf, all the F_1 tomato plants were cut leaf.
    The F_1 tomato plants were then crossed.

    What ratio of leaf types would be expected in the F_2 generation?

    A   3 potato leaf : 1 cut leaf
    B   1 potato leaf : 1 cut leaf
    C   3 cut leaf : 1 potato leaf
    D   2 cut leaf : 2 potato leaf

14. In a breed of cattle, the gene for horns (n) is recessive to the gene for no horns
    (N). True breeding cows with no horns were crossed with bulls with horns. What
    percentage of the offspring will have no horns?

    A   25%
    B   50%
    C   75%
    D   100%

15. Which of the following is a correct description of a gene?

    A   a chain of DNA bases

    B   a chain of amino acids

    C   a chain of glucose molecules

    D   a chain of phosphate molecules

16. If an inherited characteristic is controlled by alleles that have an equal effect on the phenotype of the organism, the type of inheritance is called

    A   polygenic

    B   co-dominant

    C   homozygous

    D   monohybrid

17. Biodiversity is the term used to describe different

    A   types of ecosystem

    B   genes in an organism

    C   species in an ecosystem

    D   phenotypes in an organism

18. Freshwater bony fish overcome their water balance problem by

    A   producing dilute urine

    B   excreting excess salt

    C   producing a small volume of urine

    D   drinking more water.

19. Blood entering the kidney has a urea content of 0·03%. Urine leaving the kidney has a urea content of 2·00%.

How many times greater is the urea content in the urine than in the blood?

    A   0·06

    B   1·97

    C   6·67

    D   66·67

20. Where is urea produced and from what substance?
In the

    A   kidney from amino acids

    B   kidney from glucose

    C   liver from amino acids

    D   liver from glucose

21. The table below shows daily gains and losses of water.

Water gain method	Volume (cm³)	Water loss method	Volume (cm³)
drinking	900	breathing	
eating	1000	sweating	450
metabolism	350	urine and faeces	1500

What volume of water is lost by breathing?

A   300cm³

B   550cm³

C   1050cm³

D   1950cm³

22. Digestion occurs to allow absorption of

A   large insoluble molecules

B   large soluble molecules

C   small insoluble molecules

D   small soluble molecules.

23. A student's pulse rate was measured five times at rest. The measurements were recorded in beats per minute (bpm).

Measurement	1	2	3	4	5
Pulse rate (bpm)	75	72	84	75	79

What is the average pulse rate in beats per minute?

A   75

B   77

C   192·5

D   385

Questions 24 and 25 refer to the table below which lists some blood vessels.

	Name of Blood Vessel
A	coronary artery
B	hepatic artery
C	renal vein
D	pulmonary vein

24. Which blood vessel transports blood to the liver?

25. Which blood vessel carries oxygenated blood from the lungs?

SECTION B

Try to answer all of the questions in Section B.

1.   The diagram below shows a plant cell

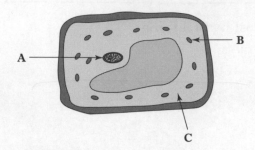

(a) (i) Name structures A and C.

A _____   **1**

C _____   **1**

(ii) Give the function of structure B.

_____   **1**

(b) Bacterial cells are used commercially.

(i) Give **two** commercial uses of bacteria.

Use 1 _____   **1**

Use 2 _____   **1**

(ii) <u>Underline</u> **one** option in each set of brackets to make the following sentence correct.

When bacteria respire aerobically they convert $\begin{Bmatrix} \text{lactose} \\ \text{glucose} \end{Bmatrix}$ sugar

into $\begin{Bmatrix} \text{carbon dioxide} \\ \text{methane} \end{Bmatrix}$ and $\begin{Bmatrix} \text{waste} \\ \text{water} \end{Bmatrix}$.   **2**

2. (*a*) Decide if each of the following statements about enzymes is **True** or **False**, and tick (✓) the appropriate box.

If the statement is **False**, write the correct word in the **Correction** box to replace the word underlined in the statement.

Statement	True	False	Correction
Enzymes are protein molecules.			
Enzymes increase the energy input required for reactions.			
Enzymes are biological substrates.			

3

(*b*) Three test tubes were set up as shown in the diagrams below, each containing a cube of solid egg white (protein) and pepsin at a different pH.

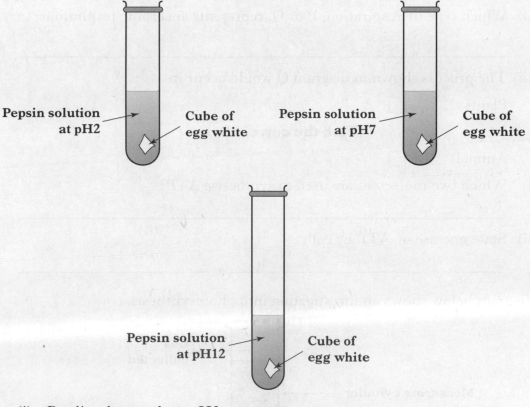

(i) Predict the result at pH2

_____

1

(ii) Name the variable altered in this experiment?

_____

1

(iii) Name two variables which should be kept constant when setting up this experiment.

_____

_____

1

3. The diagrams below show some stages in the two types of respiration.

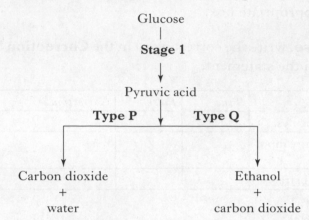

Glucose
|
Stage 1

Pyruvic acid

Type P   **Type Q**

Carbon dioxide
+
water

Ethanol
+
carbon dioxide

(a) (i) Name Stage 1.

_____ 1

(ii) Which type of respiration, P or Q, represents anaerobic respiration?

_____ 1

(iii) The process shown in diagram Q would occur in

Plants ☐

Tick the correct box

Animals ☐   1

(b) (i) Which two molecules are used to synthesise ATP?

_____ 1

(ii) State one use of ATP by cells.

_____ 1

4. The diagram below shows an investigation into photosynthesis.

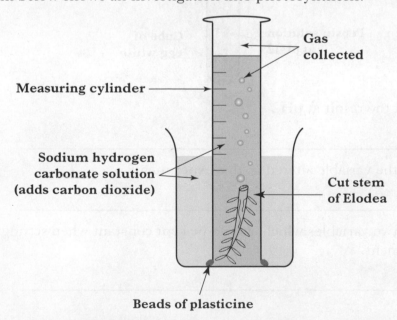

Gas collected

Measuring cylinder

Sodium hydrogen carbonate solution (adds carbon dioxide)

Cut stem of Elodea

Beads of plasticine

(a) Name the gas collected.

_____ 1

(b) The table below shows some results from this investigation.

Concentration of Sodium hydrogen carbonate solution (%)	Volume of gas collected (mm³/minute)
0·0	0·0
1·0	5·0
2·0	13·0
3·5	26·0
4·0	30·0
6·0	30·0

(i) On the grid below, plot a line graph to show the volume of gas collected against the concentration of sodium hydrogen carbonate solution.

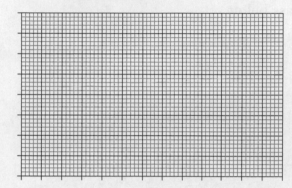

3

(ii) What is the variable being measured in this investigation?

_____

1

(iii) Name one factor that could be limiting the rate of photosynthesis between 4% and 6% sodium hydrogen carbonate concentrations of 4% and 6%.

_____

1

(iv) Suggest one way in which the reliability of these results could be improved.

_____

1

5. The diagram below shows a marine food web.

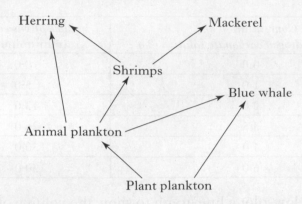

(a) Use the members of a food chain from the food web above to complete the pyramid of biomass below.

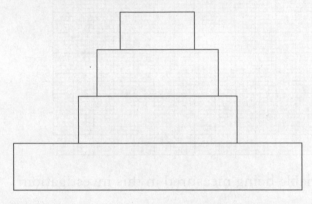

2

(b) What is represented by each arrow in the food web?

_____

1

(c) Why does the biomass decrease at each level of the pyramid?

_____

_____

1

(d) <u>Underline</u> one option in each set of brackets to make the following sentences correct.

A population consists of members of $\begin{Bmatrix} \text{different} \\ \text{the same} \end{Bmatrix}$ species. Members of a

species which eat plants and animals are called $\begin{Bmatrix} \text{herbivores} \\ \text{omnivores} \end{Bmatrix}$.

1

6. (*a*) Desert plants such as the cactus shown below are adapted to survive hot dry conditions.

Identify an adaptation of the desert plant, shown in the diagram above, for each of the following functions.

Prevents water loss   _____   **1**

Increases water uptake   _____   **1**

(*b*) Many finch species are found on the Galapagos islands as shown in the diagram below.

(i) Describe two differences in the beaks of the finches which reduce competition between species.

1. _____

2. _____   **1**

(ii) Explain why adaptations in the beaks of the finches reduce competition.

_____

_____

_____   **1**

(*c*) <u>Underline</u> one option in each set of brackets to make the following sentences correct.

The behavioural responses of woodlice $\left\{\begin{matrix}\text{increase}\\\text{decrease}\end{matrix}\right\}$ their chance of survival.

They move $\left\{\begin{matrix}\text{towards}\\\text{away from}\end{matrix}\right\}$ from hot conditions to avoid $\left\{\begin{matrix}\text{drying out}\\\text{being eaten}\end{matrix}\right\}$.   **2**

7. The diagram below shows the process of genetic engineering to produce a specific protein.

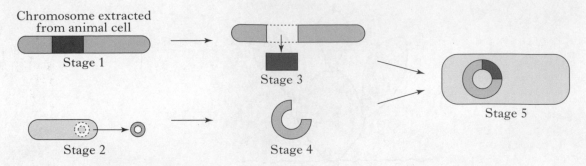

Chromosome extracted from animal cell

Stage 1

Stage 3

Stage 2

Stage 4

Stage 5

(a) Name the type of cell used in Stage 2.

_____ **1**

(b) Describe what is happening in Stages 3 and 4.

Stage 3 _____ **1**

Stage 4 _____ **1**

(c) Name a product which might be produced from the cell in Stage 5.

Give a use for this product.

Product _____ **1**

Use _____ **1**

8. Selective breeding over many generations is used to improve desirable characteristics in crop plants as shown in the table below.

Generation Number	Stem Height (cm)	Grain Yield (tonnes per hectare)
1	140	6·0
25	130	6·0
50	110	7·2
100	80	9·6

(a) Predict the grain yield in generation 110.

_____ tonnes per hectare.   1

(b) Describe the process of selective breeding.

_____

_____

_____   2

(c) Give one disadvantage of selective breeding.

_____   1

9. The diagram below shows the human alimentary canal.

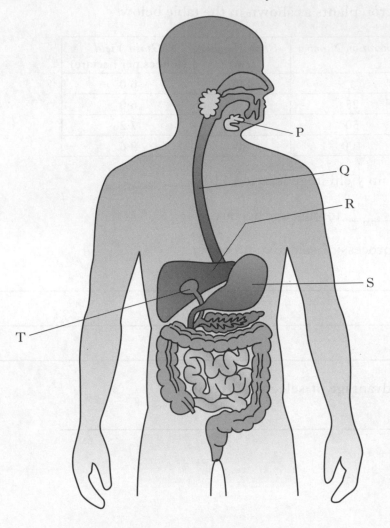

(a) Name the part labelled R in the diagram above.

_____   **1**

(b) Excess glucose is converted into an insoluble compound that is stored in the part labelled R. Name this compound.

_____   **1**

(c) Which labelled structure produces hydrochloric acid?

_____   **1**

(d) In the gut, fats and oils are emulsified by bile.
    Which label shows where bile is stored?

_____   **1**

(e) Milk is an emulsion of fat in water.

Lipase solution was added to milk and the pH of the milk was measured over 12 hours.

Time (Hours)	pH of milk
0	6·8
3	6·5
6	6·0
9	5·7
12	5·6

What was the average decrease in pH per hour?

Space for working

Average decrease in pH per hour _____   1

(f) What substance caused the change in pH of the milk?

_____   1

(g) Explain why catalase could not produce this change in pH if added to milk.

_____

_____   1

10. (a) The diagram below shows a side view of the human brain.

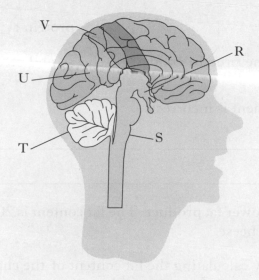

Use letters from the diagram to complete the table below.

Function	Letter
Centre of co-ordination of movement	
Control of heart rate	
Site of conscious responses	

2

(b) The following list shows stages in a reflex arc.

    A   effector produces a response

    B   receptor is stimulated

    C   impulse is carried through a motor neurone

    D   impulse is carried through a sensory neurone

    E   impulse is carried through a relay fibre

Use letters from the list to show the correct order in which these stages occur. The first stage is given.

    B → ＿＿ → ＿＿ → ＿＿ → ＿＿   **1**

(c) What is the function of reflex actions?

＿＿＿＿＿＿＿＿＿＿＿＿＿＿＿＿＿＿＿＿＿＿＿＿＿＿＿＿   **1**

11. The table below shows the composition of three different foods.

Food component	Cheese (g/100g)	Steakburger (g/80g)	Breakfast cereal (g/100g)
Protein	32·0	32·0	12·0
Fat		15·0	3·0
Carbohydrate	1·2	0	72·0

(a) What is the simple ratio of protein to fat in the breakfast cereal?

Space for working

    ＿＿ protein : ＿＿ fat   **1**

(b) A student states that the protein content of cheese and steakburger is the same.

Explain why this statement is incorrect.

＿＿＿＿＿＿＿＿＿＿＿＿＿＿＿＿＿＿＿＿＿＿＿＿＿＿＿＿＿

＿＿＿＿＿＿＿＿＿＿＿＿＿＿＿＿＿＿＿＿＿＿＿＿＿＿＿＿＿   **1**

(c) The cheese is sold as a lower fat product. The fat content is 20% less than the protein content of the cheese .

Complete the table by calculating the fat content of the cheese.

Space for working.

    1

(d) The remainder of the breakfast cereal is water.

Calculate the mass of water in the breakfast cereal.

Space for working.

＿＿＿＿＿＿ g   **1**

12. (*a*) The chart below shows changes in adult smoking patterns over the years 1935 to 2005.

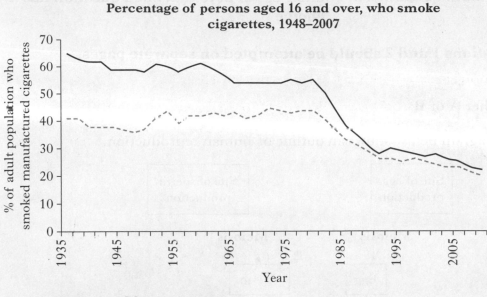

Percentage of persons aged 16 and over, who smoke cigarettes, 1948–2007

Key: — Men   --- Women

   (i)   In which year was the highest percentage of male smokers recorded?

   Year _____

1

   (ii) In which ten year period was there the greatest reduction in smoking by men?

   Tick (✓) the correct box.

   1935 – 1945 ☐
   1975 – 1985 ☐
   1985 – 1995 ☐
   1995 – 2005 ☐

1

(*b*) In 2007, the number of items prescribed by doctors to help people stop smoking was 220 000. The number increased by 25% in 2008.

How many items were prescribed in 2008?

Space for working

_____ items

1

(*c*) In the lungs, gas exchange occurs through the alveoli

   (i)   Name the process by which gases move through the alveoli walls.

   _____

1

   (ii) Give one feature of the alveoli that allows efficient gas exchange.

   _____

1

SECTION C

Try to answer both questions in Section C, but note that each question has a choice

Questions 1 and 2 should be attempted on separate pages.

1. Answer either A or B

   A. The diagram below shows an outline of human reproduction.

   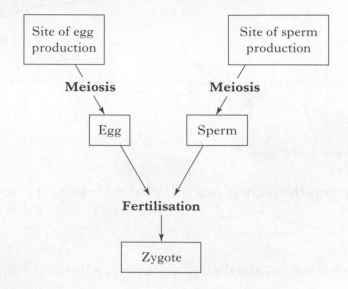

   Name the sites of production of eggs and sperm in a human. Describe how the processes of meiosis and fertilisation produce variation in humans. **5**

   OR

   B. The diagrams below shows two varieties of the Peppered moth.

   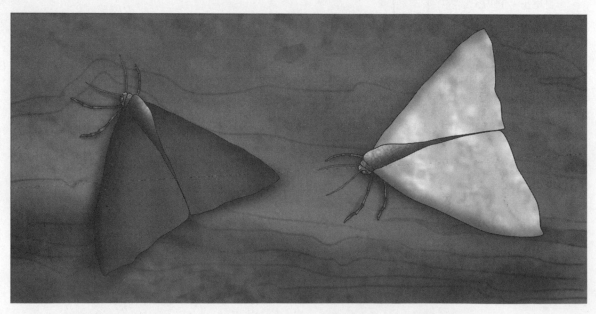

   Describe the environments that these moths are found in and the process of natural selection which causes changes in their population numbers. **5**

2.   Answer **either** A **or** B.

A. A decrease in the temperature of the surroundings is detected by the skin.
    Describe the responses that occur to maintain body temperature.   **5**

OR

B. White blood cells help defend the body against disease organisms.
    Describe the roles of white blood cells in defence.   **5**

[End of Question Paper]

Practice Exam 3

Biology   Intermediate 2

Practice Papers | Exam 3
For SQA Exams | Intermediate 2

Fill in these boxes:

Name of centre

Town

Forename(s)

Surname

You have 2 hours to complete this paper.

Instructions

Section A (25 marks)

- Complete the grid provided.

- Full instructions on next page.

Section B and C (75 marks)

- Attempt all questions and try to leave no spaces.

- Section C – two questions. Choose one option from each.

- Answer questions in ink and in any order.

- Complete graphs in pencil.

Leckie×Leckie

Scotland's leading educational publishers

Indicate your choice of answer with a single
mark in pencil as in the following example. ──────▶

	A	B	C	D
	□	■	□	□

	A	B	C	D			A	B	C	D
1	□	□	□	□		14	□	□	□	□
2	□	□	□	□		15	□	□	□	□
3	□	□	□	□		16	□	□	□	□
4	□	□	□	□		17	□	□	□	□
5	□	□	□	□		18	□	□	□	□
6	□	□	□	□		19	□	□	□	□
7	□	□	□	□		20	□	□	□	□
8	□	□	□	□		21	□	□	□	□
9	□	□	□	□		22	□	□	□	□
10	□	□	□	□		23	□	□	□	□
11	□	□	□	□		24	□	□	□	□
12	□	□	□	□		25	□	□	□	□
13	□	□	□	□						

Multiple Choice Instructions

- Check the separate grid provided and complete the details requested.

- Check that **your name** is on the answer grid.

- Read each question carefully and decide on your answer.

- Tick the correct answer on your question paper.

- Using **pencil**, transfer your answer to the separate grid.

- Shade in the box under your chosen answer.

To change an answer, erase your answer fully and shade in the new box.

Check your answers before you leave the exam room.

Leave no spaces – *all* questions should have an answer.

Make sure you only give one answer to each question.

SECTION A

Try to answer all of the questions in Section A.

Use the grid provided.

Remember to complete the grid in pencil.

1. The diagram below shows cells as seen under a microscope.

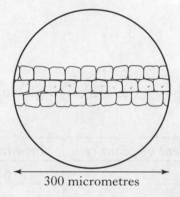

300 micrometres

The field of view is 300 micrometres across. What is the average size of one cell?

A   10 micrometres
B   30 micrometres
C   100 micrometres
D   300 micrometers

2. The structure within a plant cell that carries out photosynthesis is the

A   chlorophyll
B   cytoplasm
C   chromosome
D   chloroplast

3. Beer and wine are produced by

A   aerobic respiration by yeast
B   anaerobic respiration by yeast
C   anaerobic respiration by bacteria
D   aerobic respiration by bacteria

Questions 4 and 5 refer to the table below which shows results from an experiment into osmosis in potato cubes.

Potato cube	Starting weight (g)	Final weight (g)
P	5·0	7·0
Q	4·5	5·0
R	6·0	4·0

4. The percentage change in weight for Potato cube Q is

   A   0·5
   B   9·5
   C   10·0
   D   11·1

5. The simplest whole number ratio of starting weight to final weight in Potato cube R is

   A   6 : 4
   B   4 : 6
   C   3 : 2
   D   2 : 3

6. Which line in the table below shows the effect of placing a plant cell into a hypotonic solution?

	Change in water content of plant cell	*Appearance of plant cell*
A	gain	plasmolysed
B	gain	turgid
C	loss	plasmolysed
D	loss	turgid

7. The carbon fixation stage in photosynthesis results in the

   A   production of carbon dioxide
   B   formation of ATP
   C   formation of hydrogen
   D   production of glucose

8. Which carbohydrate is a structural component of cell walls?

   A   cellulose
   B   glucose
   C   glycogen
   D   starch

9. Which of the following describes a population within an ecosystem?
   All of

   A   the animals
   B   the plants
   C   the plants and animals
   D   one species

10. The diagram below shows part of a marine food web.

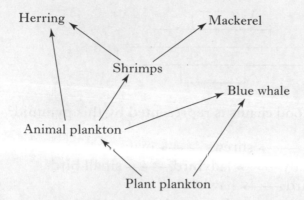

Which of the following organisms are in competition for shrimps?

A   animal plankton and herring
B   mackerel and herring
C   plant plankton and animal plankton
D   animal plankton and mackerel

11. The plant plankton can be described as

A   producers
B   primary consumers
C   secondary consumers
D   herbivores

12. Only 10% of the energy in each organism is passed to the next organism in the food chain.

In the food chain below, the animal plankton contain 500 units of energy.

plant ⟶ animal ⟶ shrimps ⟶ mackerel
plankton   plankton

How much of the energy in animal plankton is passed to the mackerel?

A   5 units
B   50 units
C   5000 units
D   50 000 units

13. The following diagram shows a pyramid of numbers.

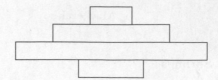

Which of the following food chains is represented by this pyramid?

A   leaves ——→ beetles ——→ shrews ——→ owl
B   oak tree ——→ greenfly ——→ ladybird ——→ small birds
C   grain ——→ small bird ——→ hawk
D   lettuce ——→ slug ——→ thrush ——→ hawk

14. The following choice chamber was used to investigate the effect of light intensity on the movement of woodlice.

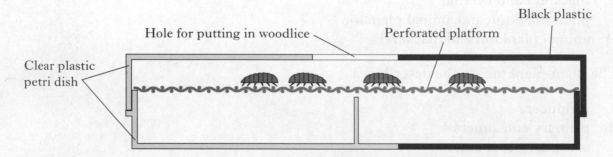

Which of the following should be the only altered variable during this investigation?

A   number of woodlice
B   temperature
C   humidity
D   light intensity

15. Farmers plant crops in a way that will yield the biggest mass of each crop. The table below shows the relationship between number of plants per square metre and mass produced for one crop.

Number of plants (per square metre)	Mass of crop produced per square metre (g)
2	300
4	600
8	800
16	1000
32	750

The reduced mass of crop when 32 plants were planted per square metre was due to

A   competition for food
B   less oxygen for photosynthesis
C   competition for minerals
D   less glucose for photosynthesis

16. Environmental impact on an organism is described by an equation. Which of the following gives the correct form of the equation?

    A   organism + genotype = phenotype
    B   phenotype + environment = genotype
    C   genotype + environment = phenotype
    D   genotype + organism = phenotype

Questions 17 and 18 refer to the table below which shows the volume of blood supplied per minute to two muscles at rest and during exercise.

Muscle	Volume of blood supplied per minute (cm³)	
	At rest	During exercise
X	100	350
Y	50	35

17. What is the percentage decrease in blood supply to muscle Y during exercise?

    A   7·5
    B   15
    C   30
    D   70

18. The volume of blood supplied to Muscle X was measured for 5 minutes at rest and for 10 minutes during exercise.

    What was the total volume of blood supplied to the muscle?

    A   4 litres
    B   600 litres
    C   2750 litres
    D   4000 litres

19. Which chamber of the heart pumps out blood that enters the pulmonary artery?

    A   right atrium
    B   right ventricle
    C   left atrium
    D   left ventricle

20. When the body is at rest, the blood entering the aorta has a pressure of 120 mmHg. This is five times higher than the pressure of the blood entering the pulmonary artery. What is the pressure of blood entering the pulmonary artery?

    A   24 mmHg
    B   95 mmHg
    C   115 mmHg
    D   600 mmHg

Questions 21 and 22 refer to the diagram of the brain shown below.

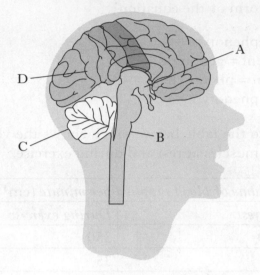

21. Which labelled structure is the cerebellum?

22. Which labelled structure detects changes in the water content of the blood?

23. Which of the following sequences shows the correct order in which an impulse passes through a reflex arc?

A   relay fibre → sensory neurone → motor neurone
B   motor neurone → relay fibre → sensory neurone
C   relay fibre → motor neurone → sensory neurone
D   sensory neurone → relay fibre → motor neurone

Questions 24 and 25 refer to the table below which shows the nutrition information on the labels of a breakfast cereal.

Ingredients	Mass per portion
Carbohydrate	66g
Protein	6g
Fat	4g
Vitamins	1·5mg
Iron	2·4mg

Component of food	Energy content (kJ per gram)
Carbohydrate	16
Protein	17
Fat	38

24. What is the energy content of one portion of the breakfast cereal?

A   71 kJ
B   76 kJ
C   437 kJ
D   1310 kJ

25. One portion of the cereal would provide 20% of a child's daily iron requirement.

How much iron is required daily by a child?

A   0·12mg
B   0·48mg
C   12mg
D   48mg

SECTION B

Try to answer all of the questions in Section B.

1. The diagram below represents part of the reproductive cycle in humans.

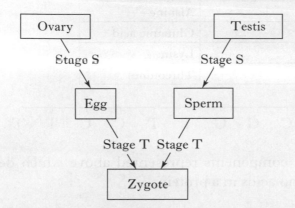

(a) (i) Name the processes occurring at stages S and T.

S ___meosis___ 1

T ___Fertillisation___ 1

(ii) Describe one process occurring in these stages that increases variation in offspring.

___random assortment of alleles which produces genetically different offspring___ 1

(b) Complete the table below to show the number of chromosomes in each of the cells.

Type of human cell	Number of chromosomes
sperm	23
liver	23
red blood cell	0

2

2. DNA contains the genetic code. Table 1 shows the DNA genetic code for some amino acids.

Table 1

DNA Genetic Code	Amino acid
CGG	Alanine
CTT	Glutamic acid
TTC	Lysine
TGA	Threonine

Section of
DNA T G A C G G T T C C T T

(a) (i) Name the DNA components represented above which determine the sequence of amino acids in a protein.

_____ 1

(ii) Using the information in table 1, complete table 2 to show the sequence of amino acids coded for by the section of DNA shown above. The first amino acid is given.

Table 2

Amino acid 1	Amino acid 2	Amino acid 3	Amino acid 4
Threonine			

2

(b) Which cell structure contains DNA?

_____nucleus_____ 1

(c) Explain the role of the amino acid sequence in protein function.

_____the amino acids assemble_____
_____the genetic code kind in_____
_____DNA_____ 1

3. Coat colour in mice is controlled by a single gene. The diagram below shows the results of a cross between a brown coated mouse and a white coated mouse.

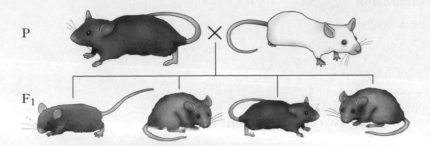

P   ×

F₁

(a) Using information from the diagram above, state which mouse coat colour is dominant. Give a reason for your answer.

Dominant coat colour ___brown___ ①

Reason ___white is recessive so the dominat gene takes over to produce brown mice___ ①

(b) Decide if each of the following statements about this cross is **True** or **False**, and tick (✓) the appropriate box.

If the statement is **False**, write the correct word in the **Correction** box to replace the word underlined in the statement.

Statement	True	False	Correction
The parents are heterozygous.	✓		
The alleles of the coat colour gene are brown and white		✓	phenotype
The genotype of the F₁ offspring is brown.			phenotype

② 3

4. The peppered moth is found in two forms, light and dark. The table below shows the number of each form in two different environments.

	Environment A	Environment B
Number of light moths	350	210
Number of dark moths	150	330

(a) Which environment is industrial? Give a reason for your answer.

Environment ___Dark Environment B___

Reason ___They prefer dark sooted coverd area in poluted think___ ①

(b) Name the process which results in the different numbers of moths in each environment.

___Natural selection___ ①

(c) What percentage of the moths in environment A were light coloured?

Space for Calculation

62.5%

_____ %

1

5. The diagrams below show two types of plant cell.

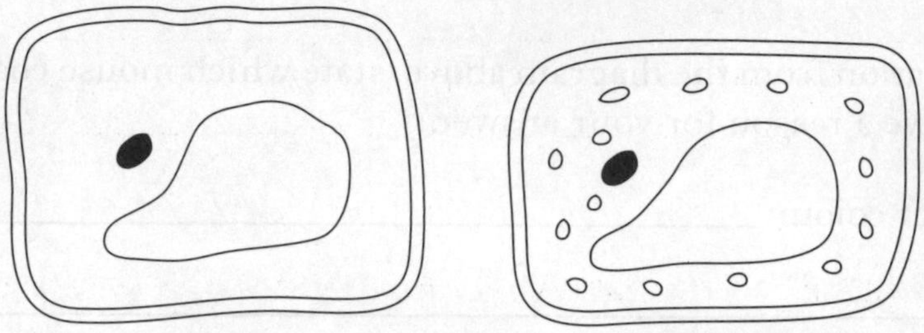

Cell type A   Cell type B

(a) Which of the cell types would be found in a green leaf?
Give a reason for your answer.

(i) Cell type ___plant___

1

(ii) Reason: ___has cell wall___

1

(b) Underline **one** option in each set of brackets to make the following sentence correct.

In bread making yeast cells respire { aerobically / anaerobically }

to produce { ethanol / carbon dioxide } which makes dough rise.

1

(c) (i) Name the type of substance, produced by fungi, which prevents the growth of bacteria.

___antibiotics___

1

(ii) Explain why these substances have no effect on some bacteria.

___some bacteria are resistant___

1

6. (a) Decide if each of the following statements about the enzyme amylase is **True** or **False**, and tick (✓) the appropriate box.

If the statement is **False**, write the correct word in the **Correction** box to replace the word <u>underlined</u> in the statement.

Statement	True	False	Correction
Amylase breaks down <u>fat</u>.		✓	~~protein~~ starch
The end product of the action of amylase in saliva is <u>glucose</u>.		✓	maltose
The action of amylase is a <u>degradation</u> reaction.		✓	enzyme

3

(b) Complete the blanks in the following equation to show the action of the enzyme phosphorylase.

$$\underline{\text{G-1-P}} \xrightarrow{\text{phosphorylase}} \underline{\qquad \text{glucose}}$$

     substrate   product   2

(c) The graph below shows the effect of increasing temperature on enzyme activity.

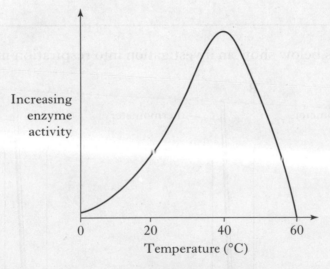

(i) Describe the effect of increasing temperature from 0°C to 40°C.

As the tem↑ the enzyme活 increases to its opt At ① 40

(ii) A. Explain why enzyme activity ceases at 60°C.

The enzyme does not work at this temp has been used up ②

B. What term is used to describe this effect?

denature ①

7. (a) The diagrams below show two anaerobic respiration pathways.

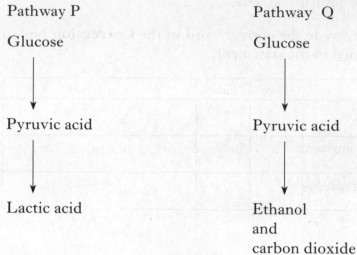

Pathway P

Glucose

↓

Pyruvic acid

↓

Lactic acid

Pathway Q

Glucose

↓

Pyruvic acid

↓

Ethanol
and
carbon dioxide

(i) Which pathway, P or Q, would occur in animals?

Pathway _____Q_____ ✗   **1**

(ii) What causes muscle fatigue in humans?

_____lactic acid_____   (**1**)

(iii) How does the body recover from muscle fatigue?

_____sweats_____   **1**

(b) The diagrams below show an investigation into respiration in yeast.

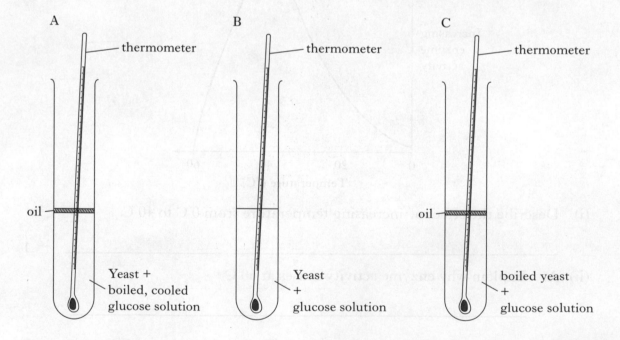

A

— thermometer

oil —

Yeast +
boiled, cooled
glucose solution

B

— thermometer

Yeast
+
glucose solution

C

— thermometer

oil —

boiled yeast
+
glucose solution

(i) In which test tube(s) will the yeast cells be respiring?

_____B_____   **1**

(ii) Why will there be no temperature rise recorded in test tube C?

_____It is boiled_____   (**1**)

8. The diagram below shows a summary of photosynthesis.

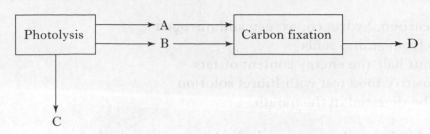

(a) Name the two substances, A and B, produced during photolysis that are used in carbon fixation.

A _____ ~~hydrogen~~ *light* 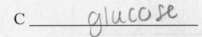   **1**

B _____ *carbon dioxide* _____   **1**

(b) Name the by-product, C, from photolysis.

C _____ *glucose* _____   **1**

(c) Product D can be converted into a storage carbohydrate.

Name this carbohydrate.

_____ *starch* _____

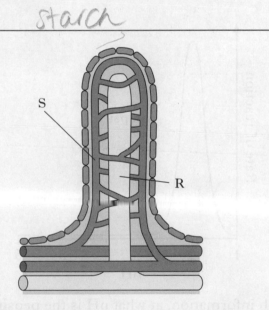

9. *(a)* The diagram below shows a villus from the alimentary canal.

(i) In which part of the alimentary canal are villi found? *small in*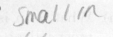

_____ *laelian body cell* _____   **1**

(ii) Name the structure labelled R.

_____ *laelian* _____   **1**

(iii) The structure labelled S absorbs food molecules.

Name one food molecule absorbed by structure S.

_____ *glucose* _____   **1**

(b) The following list contains some statements about proteins and carbohydrates.

V Contain carbon, hydrogen, oxygen and nitrogen
W Built up from amino acids
X Have about half the energy content of fats
Y Give a positive food test with Biuret solution
Z Begin to be digested in the mouth.

Complete the table below by writing all the letters from the list in the correct columns. Each letter may be used once or more than once.

Proteins	Carbohydrates
V X Y	W Z

3

(c) The graph below shows the effect of pH on the activity of the enzyme pepsin.

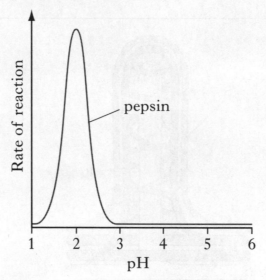

(i) From the graph information, at what pH is the pepsin most active?

pH _____2_____

1

(ii) What term is used to describe the conditions in which an enzyme is most active?

_____optimum_____

1

(iii) Pepsin acts only on proteins. It has no effect on starch. What term is used to describe this property of enzymes?

_____specific to sub_____

1

10. The following diagram represents the human circulatory system.

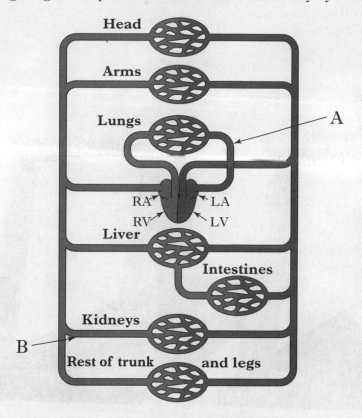

(a) Name the blood vessels labelled A and B in the above diagram.

A _____ coronary artery _____

B _____ renal artery vein _____ ① 2

(b) Describe how the structure of an artery enables it to withstand the high pressure of blood.

_____ large surrounded wall _____ ①

(c) Name the structures in veins that prevent backflow of blood.

_____ valves _____ ①

11. (*a*) The heart rates of two students were measured at rest and as they exercised on treadmills for five minutes.

The results are shown in the table below.

Time (minutes)	Heart rate (beats per minute)	
	Student A	Student B
0	70	65
1	78	67
2	84	74
3	95	80
4	102	82
5	110	84

(i) Both students A and B show an increase in their heart rates.

Calculate the difference in their heart rate increases.

Space for calculation.   40   19

Difference in increases : ___21___ beats per minute   1

(ii) Give one way in which reliability of the results would be improved.

_____repeat it_____   1

(iii) Construct a line graph of the results given in the table.

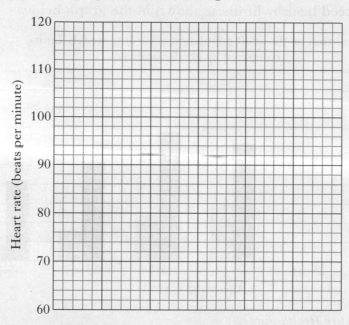

3

(b) The sentences below describe some of the features of the blood.

Underline **one** option in each set of brackets to make the sentences correct.

Oxygen is transported by $\left\{ \begin{array}{c} \text{red} \\ \text{white} \end{array} \right\}$ blood cells.

Antibodies are produced by $\left\{ \begin{array}{c} \text{macrophages} \\ \text{lymphocytes} \end{array} \right\}$.

1

(c) The figure below represents a student's breathing rate during exercise.

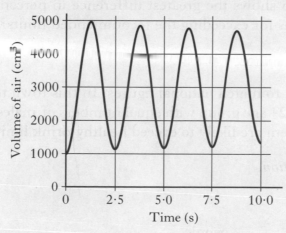

What is the student's breathing rate shown by the figure?

Space for calculation

1000 22d

_____breaths per minute.

1

12. The majority of people in Scotland drink alcohol.
Of these, many exceed healthy limits as shown in the graph below.

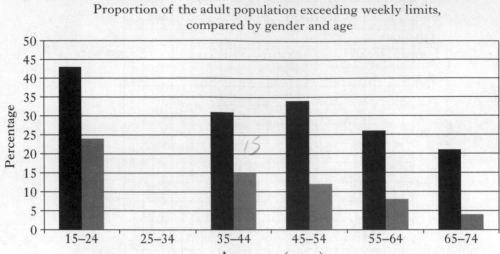

Proportion of the adult population exceeding weekly limits, compared by gender and age

(*Source: Scottish Health Survey*)

(*a*) In the age group 25–34 years, 35% of males and 17·5% of females exceed weekly limits.
Add this information to the graph above.

1

Use the graph to answer the following questions.

(*b*) (i) Which age group is least likely to exceed recommended limits?

_____ years   15 – 24

1

(ii) Which age group shows the greatest difference in percentage between males and females for exceeding the recommended limits?

_____65 – 74_____ years

1

(iii) A town is known to match national figures. In this town there are 6000 people in the 16–24 age group with equal numbers of males and females. How many of them are likely to exceed healthy drink limits?

Space for calculation.

3000 m   3000 F

6000

_____5933_____ people

1

SECTION C

Try to answer both questions in Section C but note that each question has a choice.

Questions 1 and 2 should be attempted on separate pages.

Question 1.

Answer **either** A or B

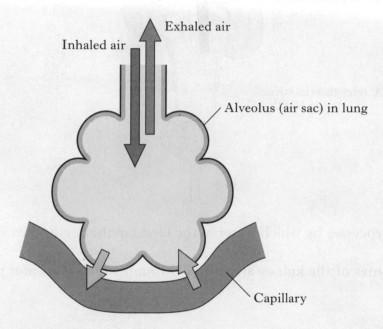

A. Describe gas exchange through the alveolus (air sac) shown above.

Include how the alveolus is adapted to allow efficient gas exchange.　　　5

The alveolus is sufficient for gas exchange due to its large surface area which allows O2 enter the cell and CO2 exit. 3 The capillary allows gas exchange to commence due to being 1 cell thick

OR

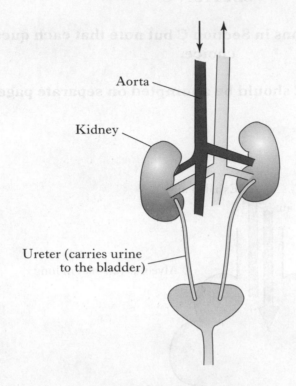

Aorta

Kidney

Ureter (carries urine to the bladder)

B. Describe the processes by which water in the blood in the aorta arrives in the ureter.
   Include the names of the kidney structures through which the water passes.

5

Question 2.

Answer **either** A **or** B

A. Describe the stages involved in the genetic engineering of bacteria to produce human insulin.

5

OR

B. Describe the different adaptations of desert plants to gain and conserve water for their survival.

5

Cactuses survive the dry desert conditions as they have long roots for water storage ③

[End of Question Paper]

Stores water in succulent tissues

Superficial roots which grow parallel to soils surface allowing the to absorb max water when rain does not fall

Worked Answers

WORKED ANSWERS   EXAM 1

SECTION A

1. D

2. B

3. C

> **HINT** Decide what antibiotics do then find bacteria under 'Effect on' column.

4. C

5. D

> **HINT** Watch the middle column of the table.

6. C

7. A

8. D

9. C

10. B

11. A

> **HINT** Underline male in question – two right answers but only one gives XY male.

12. B

> **HINT** Decide first stage – narrows your options to two answers gametes have.

13. D

> **HINT** Only one set so answer should not contain sperm or egg.

14. A

15. C

16. D

> **HINT** A not question – beware three options that will conserve water.

17. B

> **HINT** Underline plant in question – two right answers but only one for plant.

18. D

19. C

20. B

21. A

22. D

23. B

24. C

25. A

SECTION B

Answers with a '/' sign indicate correct alternatives

1. (a) hypotonic

    lower

    into   3 correct for 2 marks

                                                          2 or 1 correct for 1 mark

> **HINT** A higher % of sucrose must mean a lower water concentration.

    (b) (i) 5   1

> **HINT** Work out change in weight (16·8 − 16 = 0·8) and divide by starting weight (0·8/16) then multiply by 100.

         (ii) starting weights are different   1

> **HINT** Look at the starting weights.

         (iii) isotonic   1

2. (a) (i) Prediction – nothing / no activity   1

              Explanation – amylase does not break down hydrogen peroxide / amylase is not specific for hydrogen peroxide   1

> **HINT** Sometimes a negative result is correct.

         (ii) Control   1

> **HINT** Why would A have no enzyme added?

         (iii) Oxygen   1

    (b) Volume of oxygen produced / height of froth produced   1
         in a given time / per minute   1

> **HINT** There are 2 marks here so two points have to be given. **Rate** is an important word which suggests a time interval.

3. (a) (i) Diagram 1   1

         (ii) C   1

    (b) Heat / muscle contraction / cell division / synthesis of proteins / transmission
         of nerve impulses   Any two 2

> **HINT** These answers are taken straight from the arrangements document.

TOP EXAM TIP

Download a copy of the arrangements document p6–p28 from the SQA web site and learn the 'special' words in it.

4. (a) Chlorophyll   1

> **HINT** This question asks about a **substance** and chloroplasts would be wrong because they are **structures**.

(b) (i)   P = hydrogen   1

  Q = ATP   1

> **HINT** Look at tho direction of the arrows to see what the starting substances are.

> **TOP EXAM TIP**
> Learn the different stages in processes such as photosynthesis and both types of respiration (aerobic and anaerobic).

(ii) Carbon fixation   1

(iii) joins with carbon dioxide

  to form glucose   both   1

> **TOP EXAM TIP**
> Sometimes a bit more is expected in your answer so be confident and write that little bit more.

> **HINT** A bit of added value is wanted in this answer so give both parts.

5. (a) F   grain / lettuce   1

  F   carnivore / predator / consumer   1

  F   prey   1

> **TOP EXAM TIP**
> Question setters are not out to trap you. BUT, don't think that there will always be a mixture of true and false statements.

> **HINT** Only correct the word(s) underlined if **false**.

(b) increased

  fewer hawks to eat / prey on rabbits   both   1

> **HINT** Both parts needed to gain the mark.

(c) Community   a group of different species   1

  Population   a group of the same species   1

  Ecosystem   organisms plus environment   1

> **HINT** Four definitions: only three will be used.

6. (a) RR   WW   both   1

  R   W   both   1

> **HINT** Parents and offspring have two letters gametes have only one letter.

(b) (i) Gametes   R and W (both parents)   1

Offspring   RR, RW, RW and WW   all   1

(ii) 1 Red: 2 Pink: 1 White   1

7. (a)

3

TOP EXAM TIP

Label graph axes before doing anything else. The lack of a label is a common reason for losing an easy mark.

(b) numbers increase from 2000 in April to 18 000 in August   1

numbers decrease to 16 000 in September   1

HINT : Note 2 marks for this question – tell the **whole** story increase *and* decrease.
Give value where direction changes from increase to decrease.

TOP EXAM TIP

Averages are easy marks to get. Add up the values then divide by how many values are given.

(c) 12   1

(d) (i) 9   1

TOP EXAM TIP

For averages your answer MUST lie somewhere between the lowest and highest values. If not, then recheck your answer.

HINT : Take the values for April and August – divide August by April.

(ii) no predators /birds present to eat / prey on mud shrimps   1

HINT : Use information given at the start of the question.
Predators present August to April only.

8. (a) Bacteria   increase

Oxygen   decreases   both   1

HINT : Look at the graph and see what happens to each factor straight after the sewage input.

(b) increased bacteria use up oxygen   1

few fish can survive in low oxygen   1

TOP EXAM TIP

'Explain' questions are more difficult than 'Describe' questions. You need to **give reasons** for something happening.

HINT : This question asks you to **explain**, so not what happens but **why** it happens.

(c) the number of different species in an ecosystem   1

(d) <u>over</u>fishing (any other suitable)   1

TOP EXAM TIP

Biology uses a lot of terms. Marks are lost in exams for not being able to name parts of the body. You can learn French vocabulary so **LEARN** Biology words.

HINT : Anything that causes water pollution will not get a mark so you cannot talk about fertilisers or farm run off.

9. (*a*) (i) bronchiole   1

(ii) carbon dioxide   1

(iii) large surface area / thin walled / moist   any two   1

> **HINT** Similar properties for all absorbing surfaces (not good blood supply as this is in the stem of the question).

> **TOP EXAM TIP**
> Calculations usually are easy to do on your calculator. Be careful which buttons you press. Try to have a rough idea of the size of figure you are looking for.

(*b*) (i) 70   1

> **HINT** Decrease for P was 7, half of this is 3·5, take away from 73·5.

(ii) 10%   1

> **HINT** Decrease was 7 so $7 \div 70 \times 100$.

(iii) 200 cm³   1

> **HINT** 250 cm³ of muscle is ¼ kg so calculate ¼ of 800.

10. receptors

hypothalamus

dilate

increase   4 for 3 marks

3 for 2 marks

2 or 1 for 1 mark

11. (*a*) gain = food / metabolic water   1

loss = sweat / breath / faeces or other correct, e.g. bleeding   1

(*b*) (i) Bowman's capsule   1

(ii) Urea   1

(iii) A. ADH   1

B. pituitary gland   1

12. (*a*)

Part of blood	Function
Red blood cells	**Transport oxygen**
Plasma	Transports carbon dioxide and dissolved food
Macrophages	**Phagocytosis OR engulfing then digesting of bacteria/virus/foreign material/microbe**

3 for 2 marks

2 or 1 for 1 mark

(b) Each antibody acts on only one disease organism/foreign substance.   1

(c) 230   1

> HINT   250 × 92%

SECTION C

1. A

cells have higher concentration of solutes than surroundings   1

cells have lower concentration of water than surroundings   1

problem of gain OR influx of water   1

by osmosis OR from high <u>water</u> concentration to low <u>water</u> concentration   1

overcome by urine production OR excretion of urine   1

large volume of urine produced   1

urine very dilute   1

other point not in arrangements document e.g. salt absorbed through gills, few small glomeruli, kidney filtration rate high, little water drunk   1

Any 5 marks

> **TOP EXAM TIP**
> Write down what you can remember quickly as a list. As you write your extended response score out each point in your list.

1. B

food churned   1

muscles longitudinal/circular   1

churning mixes food with digestive/gastric juice   1

pepsin digests/breaks down protein   1

cells secrete pepsin   1

cells secrete mucus   1

mucus stops lining being digested   1

cells secrete acid   1

Any 5 marks

> **TOP EXAM TIP**
> Write your quick list. DO NOT score it out until you have completed your answer. It may still be marked if you haven't had time to finish.

2. A

Carbon dioxide (concentration)/Light intensity / Temperature   all 3 for 2 marks

2 for 1 mark

1 for 0 marks

paraffin heater / pump CO_2 in   1

add supplementary lighting   1

paraffin heater / additional heating   1

(credit paraffin heater only once)

increases crop yield   1

crops earlier   1

maximum 3

maximum = 5

2.   B

Yeast (cells) and Bacteria (cells)   both 1

The following **MUST** be linked with the correct cell to gain the mark

For yeast-

brewing / baking   1

glucose to carbon dioxide   1

and ethanol (alcohol)   1

flavourings   1

maximum 2

For bacteria-

yoghurt / cheese making   1

lactose (sugar) to lactic acid   1

souring of milk   1

any correct product from genetic engineering, e.g. insulin   1

maximum 2

maximum = 5

WORKED ANSWERS   EXAM 2

SECTION A

1. A

> **HINT** Check that the structure and function match.

2. D

> **HINT** Check all three items in the row are correct.

3. D

> **HINT** NOT question 3 correct statements plus the answer.

4. B

> **HINT** Hypertonic is a strong concentrated solution and draws water **out** of a cell.

5. C

> **HINT** Altered variable (temperature) on X axis, results always on Y axis.

6. C

7. A

8. C

9. B

10. B

> **HINT** Be familiar with all of these terms.

11. D

> **HINT** Read carefully which parents used.

12. A

> **HINT** F_2 75% dominant type; 25% recessive (ie dwarf).

13. C

> **HINT** Read carefully – cut leaf dominant.

14. D

> **HINT** Read carefully – no horns dominant.

15. A

16. B

17. C

> **HINT** Learn – often asked.

18. A

> **HINT** Learn this for one type of fish; the other is opposite.

19. D

> **HINT** Answer = 2.00 divided by 0.03.

20. C

21. A

> **HINT** Answer = total volume of water gained minus (450 + 1500).

22. D

23. B

> **HINT** Average add five numbers together and divide by 5.

24. B

> **HINT** Learn blood vessels often asked and poorly done.

25. D

TOP EXAM TIP

Use a pencil to make changes easier.

TOP EXAM TIP

Check the answer grid and leave no blanks.

TOP EXAM TIP

Tick the correct answers on the paper as you work through it.

TOP EXAM TIP

Remember to leave time to complete your grid.

SECTION B

1. (a) (i) A = nucleus   1

              C = cytoplasm   1

        (ii) B = chloroplast = site of photosynthesis   1

HINT Examine the diagram carefully and check which need structures and which functions.

    (b) (i) yogurt / cheese / insulin / biogas   any two   2

        (ii) glucose

              carbon dioxide   3 correct = 2 marks

              water   1/2 correct = 1 mark

HINT Make it clear which option you have chosen.

2. (a) True   1

              False   decrease   1

              False   catalysts   1

HINT Read each statement carefully and, if false, correct **only** the word underlined.

    (b) (i) egg white cube digested / broken up / reduced in size   1

        (ii) pH   1

        (iii) concentration/volume of pepsin

              Mass/surface area of egg white

              temperature   any 2   1

HINT For part (ii) you need to work out what changes between the test tubes. To part (iii) you need to work out what stays the same between the test tubes.

3. (a) (i) glycolysis   1
        (ii) Q   1
        (iii) tick at plants   1
    (b) (i) ADP and Pi   1
        (ii) to provide energy [for chemical reactions]/muscle contraction/
              transmission of nerve impulses/cell division   1

4. (a) oxygen   1

(b) (i)

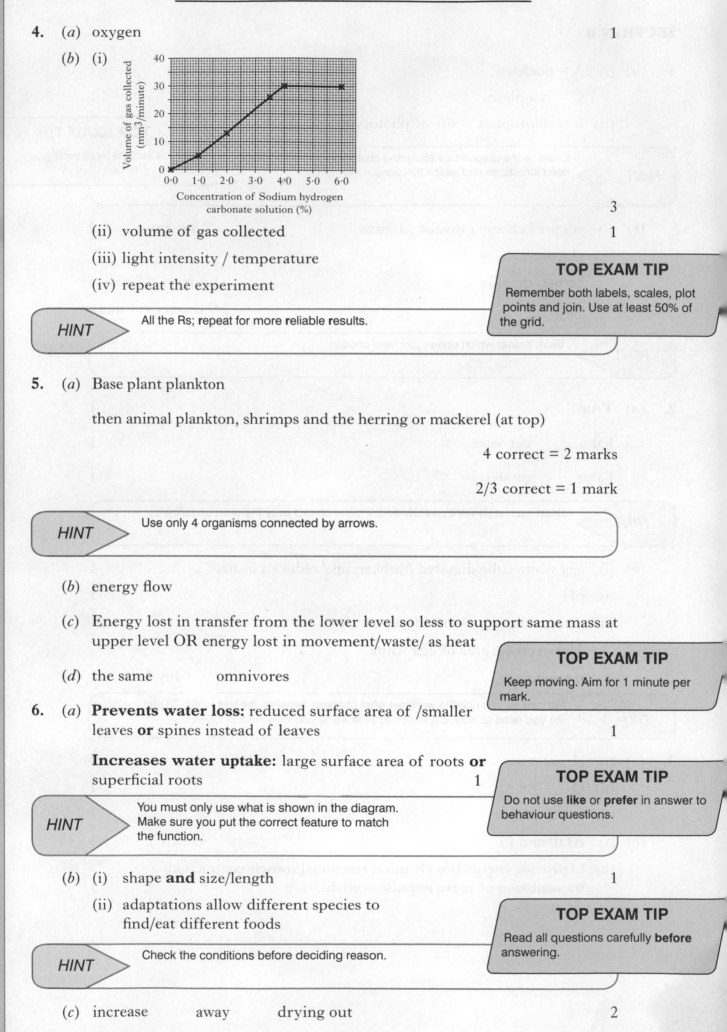

3

(ii) volume of gas collected   1

(iii) light intensity / temperature

(iv) repeat the experiment

> **HINT** All the Rs; repeat for more **r**eliable **r**esults.

> **TOP EXAM TIP**
> Remember both labels, scales, plot points and join. Use at least 50% of the grid.

5. (a) Base plant plankton

then animal plankton, shrimps and the herring or mackerel (at top)

4 correct = 2 marks

2/3 correct = 1 mark

> **HINT** Use only 4 organisms connected by arrows.

(b) energy flow

(c) Energy lost in transfer from the lower level so less to support same mass at upper level OR energy lost in movement/waste/ as heat

(d) the same   omnivores

> **TOP EXAM TIP**
> Keep moving. Aim for 1 minute per mark.

6. (a) **Prevents water loss:** reduced surface area of /smaller leaves **or** spines instead of leaves   1

Increases water uptake: large surface area of roots **or** superficial roots   1

> **HINT** You must only use what is shown in the diagram. Make sure you put the correct feature to match the function.

> **TOP EXAM TIP**
> Do not use **like** or **prefer** in answer to behaviour questions.

(b) (i) shape **and** size/length   1

(ii) adaptations allow different species to find/eat different foods

> **TOP EXAM TIP**
> Read all questions carefully **before** answering.

> **HINT** Check the conditions before deciding reason.

(c) increase   away   drying out   2

7. (*a*) bacteria 1

> HINT Learn these stages in order – often asked.

   (*b*) Stage 3   gene cut out of/removed from chromosome 1

   Stage 4   plasmid cut open 1

> HINT Look for clues in the diagram.

   (*c*) Insulin 1

   Treatment of diabetes 1

> HINT Other examples possible but this one easy to give both product and use.

8. (*a*) 9·8 (any value up to 10·5) 1

> **TOP EXAM TIP**
> Check the question for useful information.

> HINT Rises 1·2 then 2·4 only 10 years on so smaller rise.

   (*b*) organisms with desirable characteristics selected 1

   chosen organisms bred together over several generations 1

> **TOP EXAM TIP**
> Read the question and use the information.

> HINT This is a **describe** question so say **what** happens.

   (*c*) results not always guaranteed OR it takes a long time/ many generations 1

> HINT Do not give comparison with genetic engineering for this question.

9. (*a*) liver 1

   (*b*) glycogen 1

   (*c*) S or stomach 1

   (*d*) T 1

   (*e*) 0·1 units 1

> HINT Decrease is 1·2 in 12 hours so average is 0·1 units.

   (*f*) fatty acids 1

> HINT Lipase digests the fat in milk into fatty acids.

   (*g*) because catalase is not specific to/will not act on/will not fit fat **and** so fatty acids will not form 1

> HINT **Explain** so your answer must say **why** the pH not changed.

10. (a) T

    S

    U   3 correct = 2 marks; 1/2 correct = 1 mark

    (b) B → D → E → C → A   1

    (c) Rapid/ quick/protection   1

11. (a) 4 : 1   1

> **HINT**   Protein 12: Fat 3 simplified to smallest whole number ratio – divide 12 by 3.

    (b) The weight in grams in the same but the weight is per 100 g for cheese and
         per 80 g for steakburger.   1

    (c) 25·6 g   1

> **HINT**   Protein content is 32 g, 20% of this is 6·4. Fat content is 20% less than protein so fat is
>   32 – 6·4.

    (d) 13 g

> **HINT**   Total is 100 g, the components added together make up 87 g, so water is the 100 g minus
>   the 87 of other compoments.

12. (a) (i) 1935

> **HINT**   Make sure that you use the correct line for males.

         (ii) Tick at 1975 – 1985

> **HINT**   This is the steepest drop in the graph line.

    (b) 275 000

> **HINT**   25% of 220 000 is 55 000. Add 220 000 + 55 000.

    (c) (i) diffusion
         (ii) large surface area/thin walls /moist lining/good blood
               supply.   any one = 1

TOP EXAM TIP

Now go back and check you have left
no spaces. Try your best guess.

SECTION C

EXAM ANSWER ADVICE

This paper has an extended response question for Unit 2 – question 1 and unit 3 – question 2. Each paper will only cover 2 units in Section C.

You must answer two questions. One from question 1 and the other from question 2.

Choose the option which you think that you know most about.

You can use a list of bullet points/single sentences but rarely single words for this section.

For 5 marks try for 6 – 7 bullet points.

Diagrams on their own will rarely gain full marks so try to give bullet points to explain your diagram.

Remember these 10 marks can make the difference between one grade and the next or pass and fail so try them – do not leave blank.

Research has shown that if you read these options first, do the rest of the paper and then come back – marks improve. WORTH A TRY.

Read the question and underline all the main areas to be covered.

Tick each area as you cover it – full marks only if all areas covered.

1.   A

Names

Egg   ovary   maximum 1

Sperm   testes

Meiosis

Chromosomes pair

Variety of ways of pairing/random assortment of chromosomes

Increases variation in <u>gametes</u>*   maximum 2

Chromosomes divide into 4 separate gametes

Fertilisation

Nuclei of egg and sperm fuse

Many different gametes possible   maximum 2

Fertilisation is a random process

Increases variation in <u>offspring</u>*   **maximum = 5**

* mark only given once only given once – <u>underline</u> means this word must be used

B

Rural

Clean air/no pollution   maximum 1

Trees trunks light/ covered in lichen

Polluted

Soot/toxins in air   maximum 1

Trees trunks dark/ no lichens

Natural Selection

two different moths each better camouflaged in one of the environments

well camouflaged/same colour as tree trunks not seen by predators

so not eaten by predators

so survive and breed   maximum 3

numbers increase

if poor camouflage/stand out against background then eaten

so numbers decrease   **maximum = 5**

2. A

Five marks from:

negative feedback mechanism   1

blood vessels/ small arteries (not capillaries) constrict   1

reduces blood flow   1

reduces heat loss from blood   1

sweating reduced/less/stopped   1

shivering   1

repeated muscle contraction   1

generates heat   1

other correct point not in arrangements e.g. hair erected   1

                                                            maximum = 5

B

Five marks in total, maximum of 3 for each cell type.

Note that marks are only awarded if point linked to correct cell type.

Macrophages	1
Phagocytosis	1
Engulfing/ description or diagram of engulfing	1
Digestion by enzymes	1
Other correct point not in arrangements, e.g. vesicles or lysosomes	1
	maximum = 3
Lymphocyctes	1
Antibody production	1
Antibodies specific	1
Other correct point not in arrangements e.g. antigens or antibody to antigen reaction	1
	maximum = 3
	maximum = 5

WORKED ANSWERS   EXAM 3

SECTION A

1. B

> *HINT* 300 divided by 10 cells

2. D

> *HINT* Structure asked for, not substance which would be chlorophyll.

3. B

4. D

> *HINT* change ÷ original value [0·5 divided by 4·5] × 100

5. C

6. B

7. D

8. A

9. D

> *HINT* often confusion between population and community – learn them

10. B

> *HINT* look for arrows which point away from the shrimps

11. A

12. A

> *HINT* write the numbers above the chain – 10% of 500 = 50 then 10% of 50 = 5

13. B

> *HINT* unusual shape – pyramid of numbers, so low number of large producer – one oak tree

14. D

> *HINT* Altered variable is the factor that you are investigating.

TOP EXAM TIP

Work your way through the paper. If you find any question particularly difficult, do not carry on with it. Come back to it later. You will lose valuable time being bogged down with one question.

15. C

16. C

17. C

> *HINT* Decrease is 15. 15/50 × 100 = 30%

18. A

> *HINT* (5 × 100) + (10 × 350) = 4000 cm³ = 4 litres

TOP EXAM TIP

Wallpaper – Write down key terms or concepts you have to learn on small sticky notes. Post these notes around your room or house, e.g. by your bed, on the toilet door, covering the photo of you as a baby....

19. B

20. A

> *HINT* 120 ÷ 5 = 24 mmHg

TOP EXAM TIP

Get valid results by controlling variables.

21. C

22. A

> *HINT* The hypothalamus does this.

TOP EXAM TIP

Revise with a few friends: one person has the notes and asks the others questions; change round; help each other or make it a competition.

23. D

24. D

> *HINT* [66 × 16] + [6 × 17] + [4 × 38]

25. C

> *HINT* 2·4 × 5 to calculate for 100%

SECTION B

1. (a) (i) S   meiosis/gamete formation   1

           T   fertilisation   1

      (ii) in meiosis random assortment of chromosomes produces variation in gametes

            OR

            in fertilisation gametes combine randomly to produce variation in offspring   1

   (b) sperm 23

> **HINT** A gamete so 23 chromosomes (half normal cells).

        liver 46

> **HINT** A normal body cell so 46 chromosomes.

        red blood cell 0   2

> **HINT** No nucleus so 0 chromosomes.

2. (a) (i) base(s)   1

      (ii) alanine   lysine   glutamic acid

                                                      1 or 2 correct for 1 mark

> **HINT** This is a problem solving question so use the information given.
>
> You are not expected to know these facts.

   (b) nucleus/ chromosome   1

   (c) determines the protein structure needed for the function of the protein   1

> **HINT** Learn – sequence of bases in DNA determines sequence of amino acids in protein structure. This structure determines function of the protein.
>
> This is a more difficult A type question.

3. (a) brown   1

        All F_1 offspring are brown   1

   (b) False   offspring

        True

        False   phenotype   3

> **TOP EXAM TIP**
>
> **Bold print** is a cue to read the question again carefully, so you answer what is asked.

> **HINT** Do not change any words not underlined (e.g. heterozygous to homozygous).

4. (a) B

More dark moths than light   1

> **HINT**   Use the information given: do not just guess.

(b) natural selection   1

(c) 70   1

> **HINT**   Percentage part/total
> Divide the part 350 by the total 500
> Then multiply by 100 (to convert to a percentage)

5. (a) (i) B   1

(ii) chloroplasts present in B, not in A   1

(b) anaerobically

carbon dioxide   both correct   1

(c) (i) antibiotics   1

(ii) they are resistant to antibiotic /

some bacteria are only affected by a few types of antibiotics   1

6. (a) False   starch

False   maltose

True   3

> **HINT**   Do not change any words not underlined.

(b) glucose-1-phosphate   starch   2

> **HINT**   Full name of glucose – 1-phosphate needed.

(c) (i) as temperature increases, enzyme activity increases   1

(ii) A. high temperature causes a change in shape of active site
and substrate can no longer fit   2

B. denaturation/ it is denatured   1

7. (a) (i)   P   1

    (ii) build up of lactic acid (due to lack of oxygen)   1

    (iii) as oxygen becomes available the lactic acid is
      converted back to pyruvic acid/oxygen debt is repaid   1

   (b) (i)   A and B   1

> **HINT**   A anaerobic, B aerobic

TOP EXAM TIP

Read the question! This is not wasted time. Read the whole question once to got a rough idea what it is about. Then read it again from the beginning – slowly. It is a good idea to underline or highlight, including information in diagrams. Do not start until you are sure you thoroughly understand what is asked.

    (ii) yeast cells killed by
      boiling so no
      respiration/energy
      release   1

8. (a) hydrogen   1

    ATP   1

   (b) oxygen   1

   (c) starch   1

9. (a) (i)   Small intestine   1

> **HINT**   Must include the word 'small'.

    (ii) Lacteal   1

    (iii) glucose/ amino acids   1

   (b)

Proteins	Carbohydrates
V, W, X, Y	X, Z

All correct for 3 marks, three correct for 2 marks, one or two correct for 1 mark.

> **HINT**   note that X is used more than once as this is true for both.

   (c) (i)   2   1

    (ii) optimum   1

    (iii) specific/specificity   1

> **HINT**   learn these word and their meanings, they are asked in most exam papers

10. (a) A – pulmonary vein

     B – renal vein   2

   (b) thick walls/thick layer of muscle in walls   1

   (c) valves   1

11. (*a*) (i) 21   1

HINT | increase for A is 40, increase for B is 19 so subtract to find difference

(ii) repeat the investigation   1

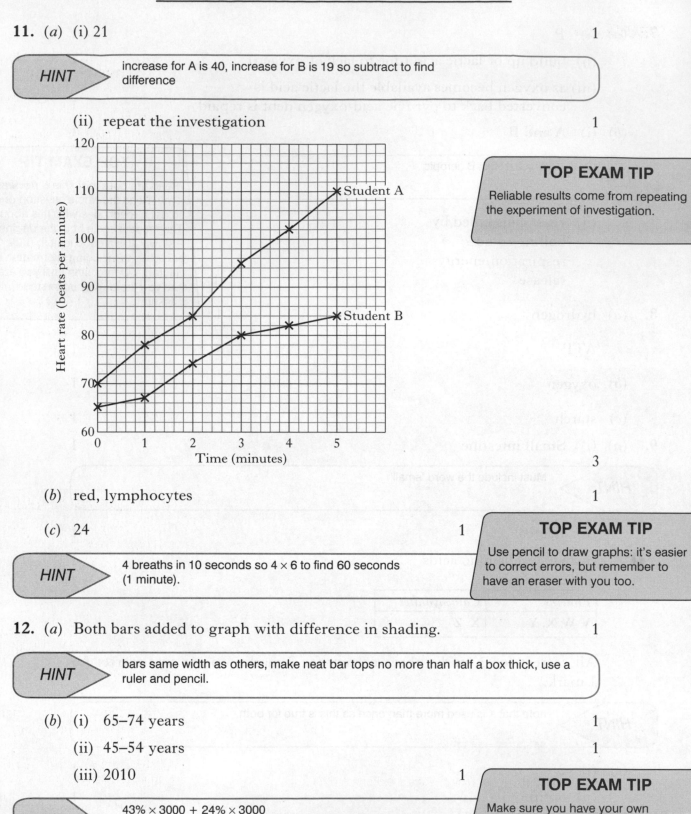

TOP EXAM TIP

Reliable results come from repeating the experiment of investigation.

3

(*b*) red, lymphocytes   1

(*c*) 24   1

HINT | 4 breaths in 10 seconds so 4 × 6 to find 60 seconds (1 minute).

TOP EXAM TIP

Use pencil to draw graphs: it's easier to correct errors, but remember to have an eraser with you too.

12. (*a*) Both bars added to graph with difference in shading.   1

HINT | bars same width as others, make neat bar tops no more than half a box thick, use a ruler and pencil.

(*b*) (i) 65–74 years   1

(ii) 45–54 years   1

(iii) 2010   1

TOP EXAM TIP

Make sure you have your own calculator and ruler for the exam.

HINT | 43% × 3000 + 24% × 3000
or 43% males and 24% females so average = 33·5% × 6000

SECTION C

1A. Five marks from:

oxygen passes into blood/from air

carbon dioxide passes out of blood/into air

by diffusion

large surface area

thin walls

moist surface/lining

good blood supply

1B. Five marks from:

renal artery

filtration

from glomerulus

into Bowman's capsule

into tubule

collecting duct

drains water out of kidney

ADH (no detail required)

2A. insulin gene identified

gene cut out of a human chromosome

plasmid removed from a bacterial cell

plasmid cut open

gene inserted into plasmid

plasmid with gene inserted into bacterial cell

use of enzymes – correctly applied to either process

bacterial cells multiply/reproduce

bacterial cells produce insulin

5

> **HINT** Learn all the stages – often asked.

TOP EXAM TIPS

1. Each paper will only cover 2 units in Section C.

2. You must answer two questions: one from question 1 and the other from question 2.

3. Choose the option for each which you think that you know most about.

TOP EXAM TIPS

Use short simple statements. Remember it's not an English essay so do not use long flowing sentences. The marks are for Biological terms and explanations.

TOP EXAM TIPS

4. Use bullet points. For 5 marks try for 6 – 7 bullet points.

5. Diagrams on their own will rarely gain full marks – add bullet points.

6. Remember essays are worth 10 marks so can make the difference between pass and fail. TRY BOTH – DO NOT LEAVE BLANK and leave early.

TOP EXAM TIPS

If you are unsure of the spelling, do your best and give it a try. You will get the mark for most words if the spelling is close to correct.

2B. Gain water

deep roots to reach deep water

roots at the surface to take in water after rain showers   maximum 3

large number/surface area of roots for water uptake

Conserve water

thick waxy cuticle decreases water loss

spines instead of leaves/reduced surface area of leaves reduces water loss

succulent tissue stores water   maximum 3

round shape reduces area for water loss

other correct point   **maximum = 5**

> **HINT**   Watch that you cover both areas of this question.

TOP EXAM TIPS

7. Research – read these section C options first, do the paper, come back – marks improve. WORTH A TRY.

8. Read the question and underline all the main areas to be covered.

9. Tick each area as you cover it – full marks only if **all** areas covered.

TOP EXAM TIPS

When you have finished do not sit with your arms folded looking up at the ceiling! You have probably not scored 100%! There are some marks still to be found. Spend the last minutes going through the paper carefully looking for errors. Answer everything: you will not get a mark for a blank space but you might squeeze a mark with a good guess.